P9-CRV-272

Grounding and Shielding Techniques in Instrumentation

Grounding and Shielding Techniques in Instrumentation

Second Edition

Ralph Morrison
Communication Mfg. Co.
Long Beach, California

A Wiley-Interscience Publication
JOHN WILEY & SONS, New York • Chichester • Brisbane • Toronto

Dedicated To Lee

Library of Congress Cataloging in Publication Data:

Morrison, Ralph.
Grounding and shielding techniques in instrumentation.

"A Wiley-Interscience publication."
Includes index.
1. Electronic instruments—Protection.
2. Electric currents—Grounding. 3. Shielding (Electricity) I. Title.

TK7878.4.M66 1977 621.381'028 77-3265
ISBN 0-471-02992-0

Printed in the United States of America

10 9 8

Preface to the First Edition

Every engineer or technician faces the problem of shielding and grounding. It is a standard part of his education to treat problems in this area. Unfortunately, his understanding is usually mixed with common sense, some "black magic," and many misconceptions. I was no exception in this educational process. For many years I worked diligently, building electronic hardware and leaving the grounding problems to relative chance. Like most of my fellow engineers, I left a few avenues open to change, took the advice of my boss, and then fussed with my project after it was built until it left my control. When things did not work out properly, it was blamed rather nebulously on ground-loops, or transformer leakage, and frequently I was unable to pin down the correct process.

I can recall trying to understand mechanisms that are now simple and routine to me. My excuses were many . . . my boss did not know the answers, I could not find them in my textbooks, and I could not accept the explanations of my fellow engineers.

Understanding comes hard—particularly when one has preconceived ideas about elementary matters. Teaching a subject does more to straighten out these misconceptions than almost any other process. This is particularly true if the individual reflects on the validity of each statement and its interrelation to other statements. In 1958 my role in life changed; I found I was in business for myself, and I had to explain why my product was better than someone else's. The principle thing I had to sell was a superior shielding technique. The explanations forced me into the role of teacher. Often, on reflection, I determined that my statements were not entirely valid or complete. This led me to a stronger statement for the next customer.

During this period it was surprising to find that I was seldom challenged, and that when I was, the challenger offered some help. Usually he too had not clearly appreciated the process or mechanism. The fact

I now often reflect on is that even though the understanding was not complete the techniques resulted in a successful company growth pattern.

I found that the biggest obstacle to my understanding was in the area of semantics. Words such as ground, earth, differential, or isolation I thought I knew thoroughly, because, after all, they were perfectly good English words. Unfortunately, they are not clearly defined words in the scientific sense and their use does not guarantee useful communication. As an example, consider the word "isolation." Loading isolation is provided by an amplifier, ground isolation is provided by a transformer, signal isolation is provided by a shield. Isolation can mean lack of ohmic continuity, isolation can mean physical separation and even emptiness. When the word "isolation" is used, I shudder, because I can only hope that I can pick up the meaning through context.

As a result of both company and product growth, the mechanisms of shielding and grounding processes became increasingly clear. Often, after discovering the explanation for a simple process, I felt that I had been blind, indeed, not to see the obvious earlier. On discussing the new point with others, I found I was teaching again.

To teach is to use words. This means communicating with terms such as ground, shield, and isolation. I found I had to define these words to use them and rarely was I challenged. Much to my surprise my definitions seemed to be acceptable. After writing this book I am left with a uneasy feeling. I am uncomfortable because I was forced to use my definitions to communicate in this area of engineering.

In order to communicate in the book, I felt obligated to use the standard jargon of the trade. I did try, whenever possible, to convert this jargon to a useful set of meanings. The result may not be generally acceptable and it may not stand the test of time, but again the attempt is to communicate mechanisms, not correct the semantic habits of the community. Time will tell how these definitions are received.

Basic physics, when properly applied, explains all known electrical phenomena. The Chapter I statement by Mr. Ohm is not in jest. Most of the shielding and grounding problems are just Ohm's law. Because the physics is so important, it was necessary to include chapters on electrostatics and magnetism. It is always necessary to keep these elementary ideas close at hand. Problems that come up are almost never caused by complex phenomena; the simple mechanisms explained by Ohm's law will serve in most cases. The reader is urged to read the first chapters with care. Their message should not be taken for granted.

The general shielding and grounding processes described in this book have been suggested to many users of instrumentation in the United States. Most of these suggestions have been tried out in practice and found to be correct. This is a comforting feeling and it adds to my confidence in writing the book.

The question I raise is, "Why hasn't someone else written this book before me?" The subject material is simple indeed. Perhaps that is the reason. The material is so simple it is glossed over as obvious. Far from that, I found it a challenge. Yes, the puzzle is simple when explained—but the problem here was to locate the puzzle. Like most engineering processes in real life, the problem is to define the problem. Given the proposition, the problem-solving process is usually easy.

I hope the material supplied in the book makes life easier for the engineer, technician, and scientist who have little time for hum and noise and ground-loops. I hope they can use it better to instrument their benches, laboratories, or specimens. I hope they can use it better to understand the instruments they own and the instruments they intend to buy. I hope it helps them to write and read specifications.

The attention given to my many in-plant seminars was much appreciated. Those engineers and instrumentation specialists who attended were an important part of my teaching experience. This experience was invaluable in setting the stage for writing of this book. Although the material took shape over many years, the seminars "jelled" the ideas into manageable form.

A thank you goes out to the many friends who offered encouragement along the way and to Dr. E. F. Guillemin, Professor (now retired) of Electrical Engineering at M.I.T., for his words of wisdom. In a few cases inadvertent remarks of instrumentation engineers (who were in trouble) spurred me on, and these people deserve my thanks. Fortunately, I was surrounded by dutiful and responsive help whenever it was needed. Mary Ann Judd typed the original manuscript. Her many hours of careful work were a blessing. Jerry McMillen's help and encouragement and her typing were greatly appreciated. Lee, my wife, deserves a real thank you for being so patient and understanding during the many evenings spent, pencil in hand. It takes many people to write a book . . . the author cannot do it alone. The understanding help of all, publisher included, are needed. Again, a thank you to all who held out a helping hand.

Ralph Morrison

MONTEREY PARK, CALIFORNIA
JULY 1967

Preface to the Second Edition

The concerns that I expressed in the preface to the first edition were apparently unnecessary. The wide acceptance of the book and its definitions has been quite gratifying.

Electronics and instrumentation have changed considerably since 1967 but the concepts of grounding and shielding have remained constant. In a few areas the new electronics is less sensitive to noise contamination, but in systems where signal conductors are physically extended, all the standard problems still need consideration. Stated another way, the physics in this area is very stable indeed.

The second edition provides a new treatment of the figures that illustrate the electrostatic enclosure. This should make the problem of visualization easier for the reader. In a two-dimensional sketch it is difficult to separate conductors from boundaries, particularly when they are often one and the same. The new figures show the shield boundaries more clearly and differentiate between enclosures at different signal potentials.

An attempt has been made to clarify areas of difficulty as they pertain to specifications. For example, shielding and common-mode rejection are closely related and a treatment of this specification was felt necessary. Notes were added to many sections to make them current. A treatment of digital grounding and system grounding was also added.

As in many first editions a few areas just did not succeed. I hope that the revision and expansion of these sections will make the reader's job a bit easier.

I express my thanks to the readers and to the publisher for making a second edition possible.

Ralph Morrison

LONG BEACH, CALIFORNIA
JUNE 1977

Table of Contents

1

Electrostatics

The electrostatic shield and the physics of electrostatic energy storage cannot be separated. As fundamental as the ideas of capacitance are, many common electrostatic effects are either ignored or not understood. The intent of this section is to review these elementary ideas, keeping in mind the need to relate these concepts to the problems at hand.

1.1 CHARGE

The basic unit of charge is the coulomb. If charge flows at a uniform rate in a conductor such that 1 C of charge passes a given point in one second, 1 A (ampere), is said to flow. The coulomb is a large unit. It takes 6.28×10^{18} electrons to make up a negative charge equal to 1 C. In practice it is usual to discuss charge in units of picocoulombs or microcoulombs. A picocoulomb is 10^{-12} C and a microcoulomb is 10^{-6} C.

1.2 FORCES BETWEEN CHARGES

Various experiments can be devised to demonstrate that charged bodies experience a force of attraction or repulsion. If an excess of electrons is placed on two bodies they are said to be charged negatively; so charged, these bodies will repel each other. If electrons are removed from two bodies they are said to be charged positively. Under this condition these bodies will repel each other. Experimentation with oppositely charged bodies shows that in this case the bodies will attract each other.

The electrostatic force f between two charged bodies is proportional to the product of their charges Q and inversely proportional to the square of their separation distance r. Expressed mathematically, without regard to units,

$$f = \frac{Q_1 Q_2}{r^2} \tag{1}$$

If the charges are of equal sign the positive product indicates a force of repulsion.

The force between two charged bodies varies as a function of the medium in which they are imbedded. These forces are greatest in a vacuum. The reduction in force is a measure of the dielectric constant k of the medium. The correct formula for the force between two charged bodies is

$$f = \frac{Q_1 Q_2}{r^2 k} \tag{2}$$

1.3 ELECTRIC FIELD

The interaction between charged bodies gives rise to the concept of electric field. A field map demonstrates the nature of the electric forces that surround a charged body. The two things that would characterize an electric field are the forces that would affect a test charge and the direction these forces would take. A test charge must be used to measure the strength of a field. This added charge must be kept very small. If this is not done the presence of the test charge will affect the character of the field being measured.

If a charge Q_1 is placed on a small sphere located in a vacuum the force field will all be directed towards the center of the sphere. Further tests can show that these forces vary inversely as the square of the distance r to the center of the sphere and proportional to the charge Q_1 on the sphere. We define the electric field E to be measured by these two quantities; therefore in a dielectric medium we have

$$E = \frac{Q_1}{kr^2} \tag{1}$$

where E is directed along r.

The force f in Eq. 1.2(2) can be written in terms of the field E in Eq. 1.3(1) by substitution:

$$f = Q_2\left(\frac{Q_1}{kr^2}\right) = Q_2 E \tag{2}$$

The expression 1.3(2) states that the field E is a direct measure of the force f, the proportionality factor being the test charge Q_2. If Q_2 is set to unity, then f and E are equal. Stated another way, *the field E is the force on a unit charge.* The vector direction of E is the same as the direction of this force.

1.4 VOLTAGE

If a charge is moved in an electric field E work must be done on the charge to effect a displacement. For example, if the force on a charge in a field is 1 dyn and the charge is moved 1 cm, then the work done on the charge is one dyn-cm or one erg. The amount of work is determined by both the field E and the charge Q involved. It is very convenient to measure work in terms of a unit charge so that a simple multiplication will yield the work for any specific charge Q.

By definition, *the work required to move a unit charge from one point to another in an electrostatic system is measured in units of volts.* The unit charge on the above definition often causes a conceptual difficulty. To actually measure the voltage between two points a small test charge would have to be used. It would have to be small enough so that the field being measured would not be modified by its presence. The work would be correspondingly small but the work per unit charge could be easily calculated.

The work required to move a unit charge between two points in an electrostatic system is analogous to the mechanical problem of lifting weights. If a 1-lb weight is raised 10 feet, 10 ft-lb of work have been expended to effect this change. We say the potential energy of the weight has been increased by 10 ft-lb. This potential energy is independent of the path taken to raise the weight to this new position.

We can parallel the above case exactly. If we do 10 V work to move a unit charge between two points, the work we do is independent of the path taken to do the work. The electric field and the gravitational field are therefore the same in this respect. A field with this characteristic is said to be a conservative field. The idea of increasing the potential energy of a weight carries over exactly in the electrostatic case. It is accepted terminology to talk about the voltage difference as being a potential difference. If a charge is moved to a point of higher potential energy its potential or voltage is said to be increased.

1.5 VOLTAGE GRADIENT

The steepness of a mountain or a hill at a given point is a measure of its potential gradient. This steepness could be measured as the ratio of potential energy increased per foot of horizontal motion. The direction of horizontal motion should correspond to a direction of steepest ascent.

In the electrical case, an electric charge Q in an electric field undergoes a change in potential energy as the charge is moved. If the direction of motion is taken along the direction of the field E, the rate at which potential energy is increased is maximum. It is therefore correct to say that the direction of maximum potential energy change is in the direction of the field E. By our previous analogy, we can measure the potential gradient by noting the change in potential energy per unit charge for a unit motion in the direction of the field. This is simply the ratio of voltage change to the distance x moved. Expressed in differential form, the gradient G is

$$G = \frac{\Delta V}{\Delta x} \tag{1}$$

where x is in the direction of E.

The term ΔV is differential voltage or differential work per unit charge. Since $\Delta V = G\Delta x$, the latter is also expressed in terms of differential work per unit charge. If G is expressed as force per unit charge, then the units are all correct, that is,

$$\Delta V = \frac{\Delta \text{ work}}{\text{unit charge}} = \frac{\text{force}}{\text{unit charge}} \cdot \Delta x = G\Delta x \tag{2}$$

G therefore equates to force per unit charge and this is exactly the definition of E given in Section 1.3. We therefore conclude that the field E is the gradient of the voltage. It is a measure of the steepness of the potential hill as seen by a charge Q. More precisely,

$$E = \frac{dV}{dx} = \text{grad } V \tag{3}$$

where x is the direction of maximum change.

1.6 A SPHERICAL CONDUCTOR WITH A CHARGE

In the previous sections the work required to move a charge in a field E was discussed. These fields were discussed without reference to practical charge distributions. Since charges must reside in or on conductors and insulators it is important to understand these arrangements in more detail.

The simplest system to consider is the single conducting sphere. When a charge Q is placed on its surface a field E exists outside of the sphere.

The field is everywhere radial and the magnitude is given by

$$E = \frac{Q}{kr^2} \tag{1}$$

The work W required to bring a small charge ΔQ from infinity up to the surface of this sphere is given by the integral of force over distance:

$$W = -\int_{\infty}^{r} \frac{Q\,\Delta Q}{kr^2}\,dr = \frac{Q\,\Delta Q}{kr} \tag{2}$$

To evaluate the potential difference between infinity and the surface of the sphere, this work must be related to a unit charge. Dividing (2) by ΔQ, we find that the potential difference is

$$\frac{W}{\Delta Q} = V_D = \frac{Q}{kr} \tag{3}$$

If the potential at infinity is defined as zero the potential at the surface of the sphere is given by 1.6(3).

It is convenient to describe the ability of this sphere to carry a charge. Since the charge and the potential are proportional from (3), the ratio of charge to voltage is dependent only on the geometry. This ratio is called the capacitance of the sphere. This ratio for a sphere is

$$C = \frac{Q}{V_D} = Q\left(\frac{kr}{Q}\right) = kr \tag{4}$$

Capacitance can be thought of as a measure of the charge carrying capability per unit of potential difference. Equation (4) simply states that a sphere of radius $2r$ has twice the charge on its surface as a sphere of radius r, assuming they are both at the same potential.

1.7 THE ELECTRIC FIELD AT A CONDUCTOR

The sphere discussed above has a field E that extends from its surface to infinity. The potential energy of this system can be thought of as being stored in the infinite field E. This idea is conceptually important, but the sphere is not representative of the geometries found in instrumentation.

Before more complex surfaces can be considered, the electric field E must be examined in more detail. Consider a conductor having an internal field E. The conductor has free charge available in the form of electrons. These electrons will experience an acceleration in the direction of the field. Because of collisions they will assume some average velocity. This is simply the statement that a current is flowing in the conductor. Because we are concerned with static processes, we have restricted ourselves to situations where all current or charge motion is zero. It is thus clear that the field E inside a conductor must be zero for a problem involving static charges. A field E can exist just at the surface but it cannot penetrate into the surface. This would imply that the charges giving rise to the field must reside at the surface only. This turns out to be the only logical conclusion one can reach. In the case of the charged conducting sphere, the charges at the surface must be uniformly distributed over the surface.

The field E must be normal to the surface of any conductor. If this were not true a tangential component of the field would exist, resulting in a surface current. Since this is disallowed in a static process, the field must leave the surface perpendicular to the surface.

1.8 THE DISPLACEMENT FIELD D

The field E due to a system of charges is reduced by the presence of dielectric material. Equation 1.6(1) gives the field E for a conducting sphere imbedded in a dielectric k. It is convenient to describe a new field D that stems only from the charges that create the field. Since both fields occur together, one can calculate D if E is known or vice versa. For the sphere of Eq. 1.6(1) the field D is given by

$$D = \frac{Q}{r^2} \tag{1}$$

Note that the dielectric constant is not present as a factor. Quite clearly then,

$$D = kE \tag{2}$$

everywhere outside the sphere. In a vacuum when $k = 1$, the displacement field and the electric field are equal, that is,

$$D = E \tag{3}$$

1.9 FIELD REPRESENTATIONS

The fields in electrostatics are analogous to the gravitational fields in mechanics. In free space these fields are continuous; that is, they exist everywhere. These fields are best represented as lines or curves that follow the direction of force on a test charge. These lines are often called lines of force or flux lines of E or D. Figure 1.9*a* shows the

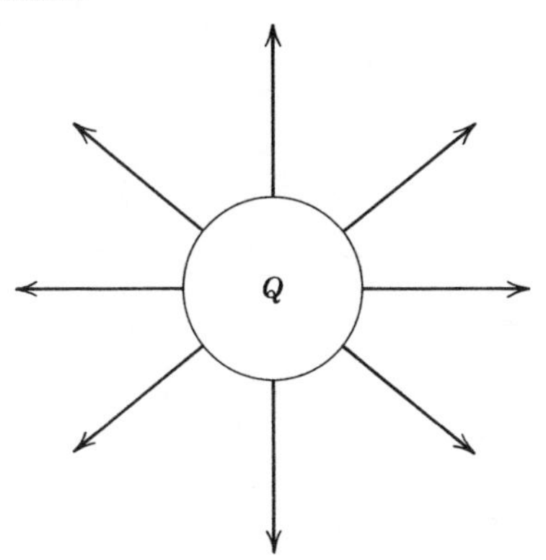

Figure 1.9*a* Lines of force representing the electric field E about a charged sphere.

radial field about a spherically charged conductor. The number of lines of force or flux lines leaving the sphere would be proportional to the charge on the sphere. Obviously this proportionality is arbitrary. One line for each unit of charge is acceptable but other ratios may be necessary to present a meaningful picture of the force field. Also, by convention, the flux lines leave position charges and terminate on negative charges.

A second method of field representation involves tubes of flux. In much the same way as lines radiate from the surface, tubes of flux, as in Figure 1.9*b*, leave the sphere at radius r_1. These tubes must cover the entire surface of the sphere and contain all the field leaving the sphere. At a larger radius r_2 these same tubes must expand to cover a larger surface area in a continuous manner. This expansion is a measure of the weakening field as the radial distance is increased. The flux density (flux per unit area) at surfaces S_1 or S_2 times the areas S_1 or S_2 must be a constant. This is another way of saying that the same flux crosses S_1 that crosses S_2.

The definition of flux that corresponds to this requirement is

$$dN = D\,dS \tag{1}$$

where dN is the element of flux crossing normal to the surface area element dS, and D is the displacement field at this point caused by the charge. An important theorem results if this definition is applied to the sphere in Figure 1.9*b*. The total value of N over the surface of the sphere can be found by integration. Since D is always normal to the surface, the integral is

$$\int_S D\,dS = 4\pi r^2 D = N \tag{2}$$

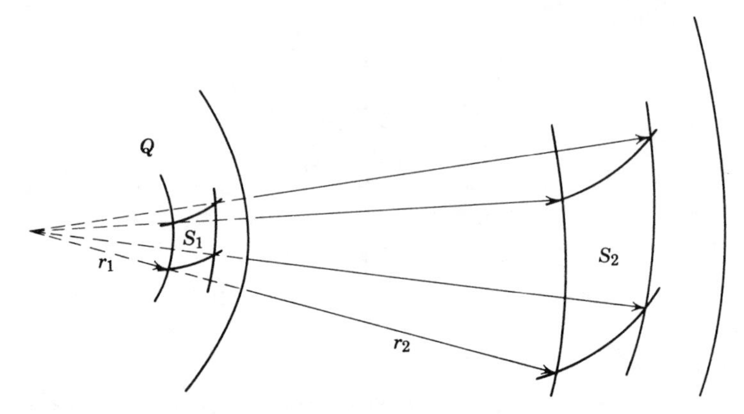

Figure 1.9b A tube of flux leaving the surface of a charged sphere.

But we know that

$$D = \frac{Q}{r^2} \tag{3}$$

Therefore we conclude that

$$N = 4\pi r^2 \left(\frac{Q}{r^2}\right) = 4\pi Q$$

This expression states that the total flux leaving a charged sphere is $4\pi Q$. This flux can be measured at any radius r. We also note that the flux is proportional to the charge Q as required; that is, the flux exists only if the charge exists. The ideas of flux leaving a sphere can be extended to flux leaving any region containing charge. Stated in the most general terms

$$\int_S D \cdot dS = 4\pi Q \tag{4}$$

where S is any closed surface and the dot product is observed for all D and S. This statement is known as Gauss' electric-flux theorem.

1.10 POINTS OF DIFFICULTY

The sphere in the previous section is convenient for a discussion of fundamental concepts without undue mathematical complexity. In other geometries, the charge distributions may not be uniform or the fields may not be mathematically simple, but the ideas remain constant.

Conceptual difficulties do arise and they should be mentioned. The sphere with charge Q cannot be reduced to radius zero without the field E going to infinity. In any practical arrangement, charges reside on at least microscopic volumes. The point charge causes no problem if the physical significance of an infinite E field is ignored. A charge distributed on the surface of a sphere develops a field that is indistinguishable from the field developed by a point charge located at the center of the sphere. The existence of this point charge is mathematical. This charge is actually not at the center and the field E does not have an infinite value.

Another difficulty involves the idea of the field D extending to infinity. If this flux emerges from positive charges it must terminate on negative charges at this infinite distance. The field at infinity is a mathematical requirement. At great distances the field strength and any charge density both reduce to zero.

In practice, all charges reside on or near the earth, and the earth is a part of the electrostatic geometry. The earth can be thought of as an infinite reservoir of charge always capable of receiving the flux from any charged body. Thus the earth in most cases represents the practical "infinity" for field D emerging from any charged body. A charged body far removed from other bodies or earth is "infinitely" removed for all practical purposes. The ideas presented are thus valid and yield meaningful insight into the character of the electric field.

The smallest charged particle is the electron. The infinitesimal charges required in the mathematics do not actually exist. The practical problems that will be considered involve charge accumulations that are large compared to a single electron. Even though these charge distributions are not continuous, the concepts derived by assuming the contrary are extremely accurate and obviously simpler in form.

1.11 MKS SYSTEM OF UNITS

Gauss' flux theorem in Section 1.10 contains a factor of 4π. It is desirable to select units such that this constant is not present in this equation. With the factor 4π removed, flux and charge are equivalent. We thus have

$$N = \int_S D \cdot dS = Q \tag{1}$$

This equality can be created by changing Eq. 1.2(2) to have the factor 4π in its denominator. If MKS units (meters, kilograms, seconds) are used and charge is expressed in coulombs the correct proportionality can be absorbed in a new dielectric factor ϵ. We thus have

$$\boxed{f = \frac{Q_1 Q_2}{4\pi\epsilon r^2}} \tag{2}$$

If we follow this arrangement the field E for a point charge is given by

$$\boxed{E = \frac{Q}{4\pi\epsilon r^2} \quad \text{and} \quad D = \epsilon E} \tag{3), (4}$$

The potential at a distance r from a charged sphere is

$$\boxed{V = \frac{Q}{4\pi\epsilon r}} \tag{5}$$

and the capacitance of the sphere is

$$\boxed{C = 4\pi\epsilon r} \tag{6}$$

Obviously this new dielectric factor ϵ is not unity in a vacuum. If ϵ_v represents the dielectric constant in a vacuum then the ratio of ϵ in a medium to that in a vacuum equals the factor k used in 1.2(2). The factor k is correctly termed the *relative capacitivity* or *relative dielectric constant* and is equal to

$$k = \frac{\epsilon}{\epsilon_v} \tag{7}$$

Equations (3) through (6) are valid in any dielectric medium where $\epsilon = \epsilon_v k$. All that remains is to establish the value of ϵ_v.[1] In the MKS system (meter–kilogram–second) the value of ϵ_v is given by Eq. (8). When Q is expressed in coulombs, r in meters, f in kilograms, and E in volts per meter, the *capacitivity of free space* ϵ_v has the value

$$\epsilon_v = 8.85 \times 10^{-12} \text{ F/m} \tag{8}$$

1.12 CHARGES ON SPHERICAL SHELLS

When a charge Q is placed on a spherical surface the potential difference between infinity and the surface of the sphere is given by 1.11(5) as

$$V_D = \frac{Q}{4\pi\epsilon r} \tag{1}$$

[1]A complete treatment of electrical units can be found in any standard text on electrostatics. The intent here is to demonstrate basic ideas and supply the elementary equations in usable form.

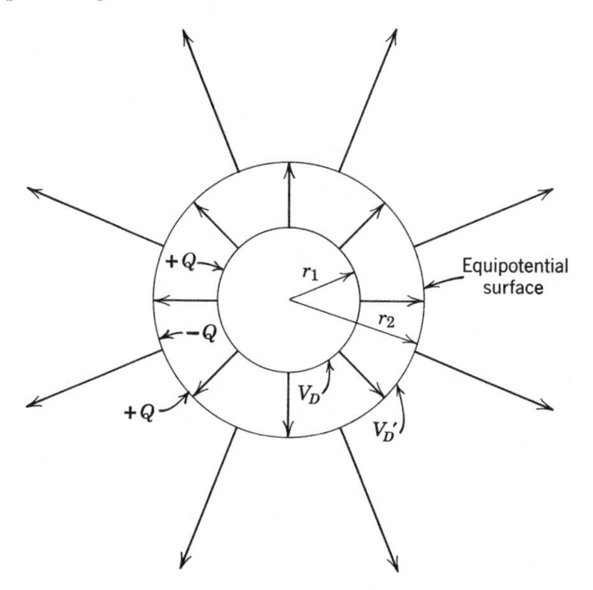

Figure 1.12*a* An equipotential surface.

The potential difference at any other radial distance greater than r say r_1 is given by

$$V_D' = \frac{Q}{4\pi\epsilon r_1} \tag{2}$$

The surface described by the new radius r_1 is spherical and is called an equipotential surface. All points at this distance r_1 are at the same potential relative to infinity or relative to the potential at the surface of the sphere.

An equipotential surface such as the sphere of radius r_1 in Figure 1.12*a* can be replaced by a thin perfect conductor without affecting the nature of the field. This assumes that the potential on the conductor is adjusted to V_D', the value before the substitution was made. It is of interest to examine the nature of this statement in more detail. The flux leaving the surface at radius r_1 due to charge Q must intercept the inside of the added surface r_2. Since this flux cannot penetrate through the conductor, a charge $-Q$ must reside on this inner surface. This must follow because flux can only start or stop on charges. This leaves a possible contradiction because no new charges were added to the system, only a conducting surface. This difficulty is overcome if a charge $+Q$ exists on the outside of the added sphere. This must be the case if the field for distances greater than r_2 is to be correct. The sum of charges residing on the inner

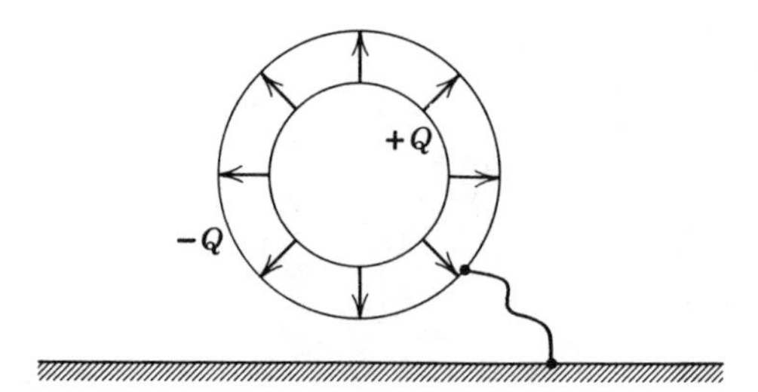

Figure 1.12b Inducing charge on a conductor.

and outer surfaces of the conductor is now zero. All the conditions are met and the field is the same as before the conductor was added. If the field external to the outside conducting surface is ignored, the field between the two spheres can be examined in more detail. The potential difference between the two surfaces is given by subtracting (2) from (1) or

$$V = V_D - V_D' = \frac{Q}{4\pi\epsilon}\left(\frac{1}{r_1} - \frac{1}{r_2}\right) \tag{3}$$

The charge stored on the inside of r_2 and on the outside of r_1 is Q. The capacitance of this system is the ratio of this charge to the potential differences or

$$C = \frac{Q}{V} = \frac{4\pi\epsilon r_1 r_2}{r_2 - r_1} \tag{4}$$

where ϵ is the capacitivity of the dielectric between the two spheres.

If a fine conducting wire is momentarily tied between the outer conducting sphere r_2 and an infinite conducting surface located nearby, the potential difference between the sphere and the conducting surface will reduce to zero. See Figure 1.12b. If a potential difference did exist, a continuous current would flow in the wire. An isolated conductor cannot support any current except for a very short period of time. After a transient current flow takes place the wire can be removed. At this time the two conductors are at the same electrostatic potential. Stated another way, no work is required to take a test charge from one surface to the other over any path. This means that no electric field exists outside the sphere r_2. This must imply that the charge on the outside surface of the sphere r_2 is zero. Since a charge $-Q$ exists on the inside surface, the conducting sphere has been charged to $-Q$ by this process. This charge is said to be an induced charge.

The Earth Plane

The infinite conducting plane shown in Figure 1.12*b* is often called an earth or ground[1] plane. For reasons of clarity we shall refer to it as the earth plane. It is usually assigned a potential of zero and all other potentials are referenced to this zero. It will be pointed out in later chapters that this is not true in practical cases. For purposes of further discussion in electrostatics, the earth plane is at zero potential everywhere and serves as an infinite source of charge. The charge $-Q$ in Figure 1.12*b* was taken from the earth conductor without changing its zero of potential.

1.13 TYPICAL CHARGE DISTRIBUTIONS

If two conducting spheres are charged oppositely, as in Figure 1.13*a*, the flux leaving one sphere terminates completely on the second sphere. The flux arriving at infinity is zero. If two conducting spheres are both equally charged, the resulting field is shown in Figure 1.13*b*. In this case the field at a large distance is the same as a single sphere of charge $2Q$. All the flux emerging from the two spheres terminates at infinity.

1.14 CYLINDRICAL SURFACES

Circular cylindrical geometries occur frequently in electronics. The fields are very similar in appearance to those in Figures 1.12*a*, 1.12*b*, 1.13*a*, and 1.13*b*. For mathematical simplicity the surfaces are considered infinite in extent and the electrical properties are treated on a per-unit-length basis.

Concentric circular cylinders with a charge Q per unit length and radius r_1 and r_2 as in Figure 1.14 have a field D confined between the two surfaces. The field D leaving surface r_1 stems from charge $+Q$. Gauss' flux theorem applied to a unit cylindrical length requires

$$\int_S D \cdot dS = Q \tag{1}$$

For any radius r, D is constant and the integral reduces to

$$2\pi r D = Q \tag{2}$$

[1] The term ground also appears in the subsequent text. Its meaning is somewhat similar to the term earth although less specific. A ground may be a large conductor or a reference element and it may or may not be ohmically connected to earth. In almost every case it is reactively associated with earth.

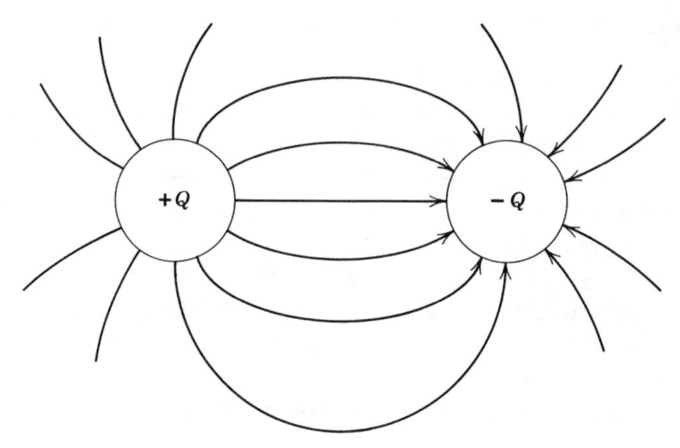

Figure 1.13a Electric field of two oppositely charged conducting spheres.

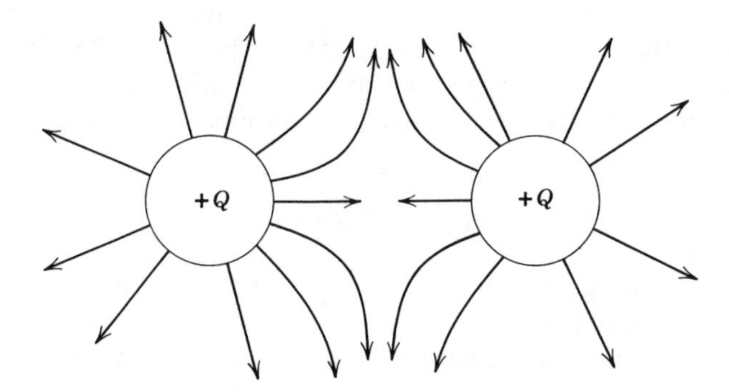

Figure 1.13b The electric field of two equally charged conducting spheres.

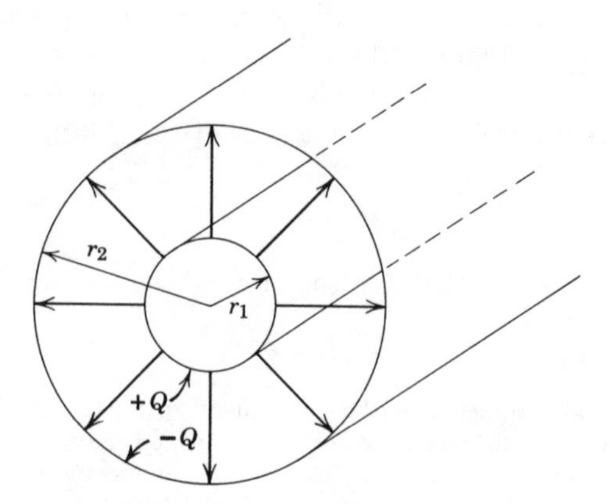

Figure 1.14 Concentric circular cylinders.

Recalling $D = \epsilon E$, the field E is thus

$$E = \frac{Q}{2\pi\epsilon r} \tag{3}$$

The potential difference between the two cylinders can be derived by integrating the work required to move a unit charge in the field between the two surfaces.

$$V_D = -\frac{Q}{2\pi\epsilon}\int_{r_1}^{r_2}\frac{dr}{r} = -\frac{Q}{2\pi\epsilon}\ln\frac{r_1}{r_2} \tag{4}$$

The capacitance per unit length is given by the ratio

$$\frac{Q}{V_D} = C = \frac{2\pi\epsilon}{\ln r_2/r_1} \tag{5}$$

These equations consider the potential difference between two circular cylinders only. Unlike the sphere, the potential in the vicinity of an infinitely long cylinder (when $r_2 \to \infty$) turns out to be infinite and physically has no meaning.

1.15 PARALLEL PLATE CAPACITORS

Assume a charge density of Q per unit area in the geometry of Figure 1.15. The displacement field emerging from the top surface is simply Q. The electric field E is given by

$$E = \frac{Q}{\epsilon} \tag{1}$$

where ϵ is the capacitivity of the dielectric between the plates. The potential difference is Ed

or

$$V_D = \frac{Qd}{\epsilon} \tag{2}$$

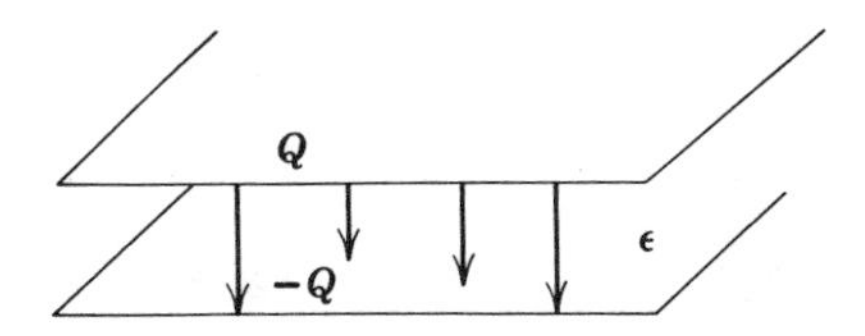

Figure 1.15 Parallel plate capacitor.

The capacitance per unit area is

$$C = \frac{\epsilon}{d} \tag{3}$$

The capacitance for any area A is therefore

$$C = \frac{A\epsilon}{d} \tag{4}$$

2

Capacitance and Energy Storage

2.1 GENERAL COMMENTS

The geometry of a group of conductors determines its energy-storing capability. This energy may be stored with some conductors charged and others discharged, or it may be stored with some conductors grounded and others raised to various potentials. These storage capabilities are measured in terms of capacitances and elastances. A discussion of these concepts logically follows from a treatment of Green's reciprocation theorem.

2.2 GREEN'S RECIPROCATION THEOREM

The potential V_1 at a distance r_1 from a charge Q_1 is given by 1.11(5) as

$$V_1 = \frac{Q_1}{4\pi\epsilon r_1} \tag{1}$$

Since this potential is just a numerical value (a scalar) a second charge at a second distance r_2 would add to this potential directly. The new potential V_{1+2} is

$$V_{1+2} = \frac{1}{4\pi\epsilon}\left(\frac{Q_1}{r_1} + \frac{Q_2}{r_2}\right) \tag{2}$$

Consider a system of charged conductors numbered 1, 2, 3, . . . , n. The potential on conductor 1 due to charges on the others is given by

$$V_1 = 0 + \frac{Q_2}{4\pi\epsilon r_{21}} + \frac{Q_3}{4\pi\epsilon r_{31}} + \frac{Q_4}{4\pi\epsilon r_{41}} \tag{3}$$

Similarly, the potentials on conductor 2 due to charges on 1, 3, 4 is given by

$$V_2 = \frac{Q_1}{4\pi\epsilon r_{12}} + 0 + \frac{Q_3}{4\pi\epsilon r_{32}} + \frac{Q_4}{4\pi\epsilon r_{42}} + \cdots \tag{4}$$

and similarly, on conductor 3,

$$V_3 = \frac{Q_1}{4\pi\epsilon r_{13}} + \frac{Q_2}{4\pi\epsilon r_{23}} + 0 + \frac{Q_4}{4\pi\epsilon r_{43}} + \cdots \tag{5}$$

If V_1, V_2, V_3, etc., are multiplied by values Q_1', Q_2', Q_3', the series of equalities that result are

$$\begin{aligned} V_1Q_1' &= 0 + \frac{Q_1'Q_2}{4\pi\epsilon r_{21}} + \frac{Q_1'Q_3}{4\pi\epsilon r_{31}} + \frac{Q_1'Q_4}{4\pi\epsilon r_{41}} + \cdots \\ V_2Q_2' &= \frac{Q_2'Q_1}{4\pi\epsilon r_{12}} + 0 + \frac{Q_2'Q_3}{4\pi\epsilon r_{32}} + \frac{Q_2'Q_4}{4\pi\epsilon r_{42}} + \cdots \\ V_3Q_3' &= \frac{Q_3'Q_1}{4\pi\epsilon r_{13}} + \frac{Q_3'Q_2}{4\pi\epsilon r_{23}} + 0 + \frac{Q_3'Q_4}{4\pi\epsilon r_{43}} + \cdots \\ &\;\;\vdots \end{aligned} \tag{6}$$

Potentials in Eqs. (3), (4), and (5) can also be written for a different set of charges Q_1', Q_2', Q_3' $\cdots$. These potentials are V_1', V_2', V_3', ... When these equalities are multiplied by Q_1, Q_2, Q_3, . . . , the statements that result are the same as (6) except that primed and unprimed values are interchanged. Note that the distances r_{21}, r_{31}, r_{41}, etc., are equal to the distances r_{12}, r_{13}, r_{14}, etc.

An examination of the sum $V_1Q_1' + V_2Q_2' + V_3Q_3' + \cdots$ can then be compared with a new sum $V_1'Q_1 + V_2'Q_2 + V_3'Q_3 + \cdots$. [These sums are formed by adding all the columns in (6) together.] Because product terms like $Q_1'Q_2$ occur in all possible subscript combinations, these two sums must be equal. Stated more precisely,

$$\Sigma V_n'Q_n = \Sigma V_nQ_n' \tag{7}$$

This is known as Green's reciprocation theorem. This theorem is a very useful tool in the field of electrostatics. Problems involving the superposition of fields or of induced charges can be treated rather elegantly using this theorem. It is discussed in this text because it serves as the basis for introducing mutual capacitance. These capacitances exist physically and they are important in instrumentation. They are never on a parts list, and therefore they are not too well understood.

2.3 SELF- AND MUTUAL ELASTANCE

An important application of Green's reciprocation theorem establishes the ratio of charge to voltage on separate conductors. Consider a system of conductors uncharged and unearthed. Assume that a charge, when placed on conductor 1 only, causes a potential V_2 on conductor 2; and a charge Q_2' when placed on conductor 2 only, causes a potential V_1' on conductor 1. We desire to show that

$$\frac{V_1'}{Q_2'} = \frac{V_2}{Q_1} \tag{1}$$

This statement is proved by using Green's theorem, which requires

$$V_1'Q_1 + V_2'Q_2 = V_1Q_1' + V_2Q_2' \tag{2}$$

We are given that Q_1 develops V_2 with $Q_2 = 0$ and that Q_2' develops V_1' with $Q_1' = 0$. Combining these statements into (2) we see that

$$V_1'Q_1 = V_2Q_2' \tag{3}$$

which is the same statement as (1).

The ratio (1) is called the coefficient of potential or mutual elastance and is usually given the symbol s_{12}:

$$s_{12} = \frac{V_2}{Q_1} \quad \text{and} \quad s_{21} = \frac{V_1}{Q_2} \tag{4}$$

Statement (3) proves that $s_{12} = s_{21}$. (5)

Coefficients of mutual elastance can be found for each pair of conductors. These coefficients are functions of the geometry only as they are ratios of voltage to charge. These coefficients are always positive as a single positive charge raises the potential everywhere. By superposition the potential on each conductor can be expressed as a sum of these coefficients, i.e.:

$$\begin{aligned} V_1 &= s_{11}Q_1 + s_{21}Q_2 + s_{31}Q_3 + \cdots \\ V_2 &= s_{21}Q_1 + s_{22}Q_2 + s_{32}Q_3 + \cdots \\ V_3 &= s_{13}Q_1 + s_{23}Q_2 + s_{33}Q_3 + \cdots \end{aligned} \tag{6}$$

Terms like s_{11}, s_{22} are called self-elastances and are not to be taken as reciprocals of self-capacitances. The ratio V_1/Q_1 equals s_{11} when Q_2, Q_3, Q_4, etc., are zero. This implies that these conductors are unearthed to avoid induced charges. This is not the condition used in defining self-capacitance.

2.4 SELF- AND MUTUAL CAPACITANCE

The inversion of 2.3(6) provides an expression for charge as a function of element potentials. The new coefficients are algebraic combinations of the elastance coefficients. The charges are

$$\begin{aligned} Q_1 &= c_{11}V_1 + c_{12}V_2 + c_{13}V_3 + \cdots \\ Q_2 &= c_{21}V_1 + c_{22}V_2 + c_{23}V_3 + \cdots \\ Q_3 &= c_{31}V_1 + c_{32}V_2 + c_{33}V_3 + \cdots \\ &\cdot \\ &\cdot \\ &\cdot \end{aligned} \tag{1}$$

Terms like c_{12}, c_{31}, etc., are called mutual capacitances. Terms like c_{11}, c_{22}, etc., are called self-capacitances.

The mutual capacitance c_{12} is the ratio of induced charge Q_1 to the potential V_2 with V_1, V_3, V_4, V_5, etc., equal to zero. (This is equivalent to saying that these conductors are earthed.) Because induced charges are of opposite sign, all mutual-capacitance terms are negative. Note that self-capacitance terms are always positive. Green's theorem can again be used to show that $c_{12} = c_{21}$, $c_{31} = c_{13}$, and so on.

Although mutual-capacitance terms can be derived from elastance coefficients, it is obvious that they are not simply reciprocals of each other. In most practical problems mutual capacitances are small enough to be neglected. When they are important they are very important and more will be said about them in subsequent chapters.

2.5 ELECTRIC SCREENING (SHIELDING)

The concept of mutual capacitance can perhaps be made clearer if one considers the system of conductors in Figure 2.5*a*. If conductor ① has a charge Q_1 then conductor ②, which surrounds ①, must have a charge $-Q_1$ because all the flux from ① must terminate on ②. The external field about ② and ③ is zero; therefore Q_3 equals

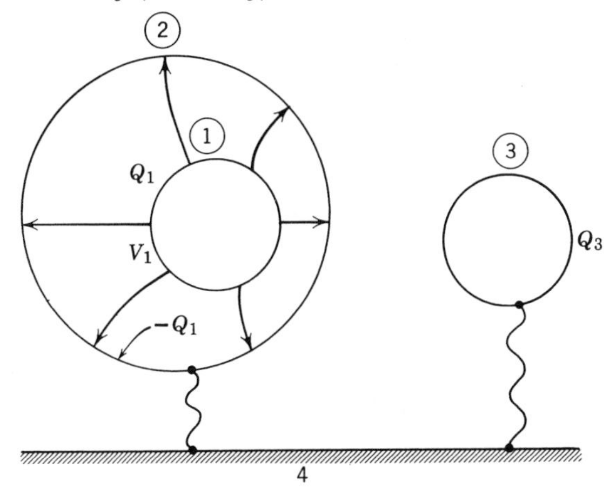

Figure 2.5*a* A system with electric screening of conductor ②.

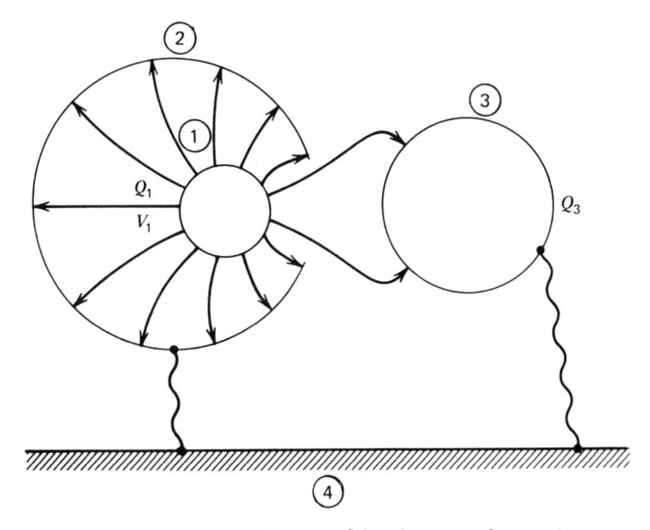

Figure 2.5*b* Leakage flux resulting in mutual capacitance.

zero. The ratio of $Q_3/V_1 = c_{31} = 0$. Conductor ① is said to be completely screened.

Figure 2.5*b* shows a system with mutual capacitance. Here the flux leaving ① does not terminate entirely on ②. Some flux escapes to terminate on ③ resulting in an induced charge Q_3. The ratio Q_3/V_1 is negative and is the mutual capacitance term c_{31}. Note that $|Q_3 + Q_2| = Q_1$ if we assume flux from ① terminates on ② and ③ only and not on the earth plane ④.

2.6 ENERGY IN A SINGLE CAPACITOR

The energy stored in a system is equal to the work required to place the charges into the system. As has been shown, in a practical capacitor C, which usually consists of two closely spaced conductors, one conductor is charged positively with a charge Q, and the second conductor has an equal negative charge on its surface. When the capacitor is fully discharged no work is required to move a first incremental charge from one conductor to the other. When the second increment of charge is moved across, work must be done against the field E caused by the first incremental charge. This work is just the voltage between the two plates times the second incremental charge. After a total charge Q has been accumulated, the work required to bring the next increment of charge is $V\,\Delta Q$. This increment of work ΔW is

$$\Delta W = V\,\Delta Q = \frac{Q}{C}\,\Delta Q \tag{2}$$

The total work for all charges ΔQ is just the integral of (2) from 0 to Q. Passing to the integral to sum the work,

$$W = \int_0^Q \frac{Q\,dQ}{C} = \frac{Q^2}{2C} \tag{3}$$

This can be expressed in terms of voltage by using the definition $C = QV$. Thus we have

$$W = \tfrac{1}{2}CV^2 \tag{4}$$

$$W = \tfrac{1}{2}QV \tag{5}$$

2.7 ENERGY STORED ON A MULTIPLE-CONDUCTOR SYSTEM

Sections 2.7 and 2.8 are supplied for completeness only. They are not required for an understanding of subsequent material. The arguments of Section 2.6 can be applied to every conductor pair in a system. The energy stored on each conductor m is $\frac{1}{2}V_mQ_m$. The total energy is therefore

$$W = \tfrac{1}{2}\sum^{m} V_mQ_m \tag{1}$$

If Eq. 2.3(6) or 2.4(1) is substituted into 2.7(1) the energy can be written in terms of elastances or capacitances

$$W = \tfrac{1}{2}(s_{11}Q_1^2 + 2s_{12}Q_1Q_2 + s_{22}Q_2^2 + 2s_{13}Q_1Q_3 + \cdots) \tag{2}$$

$$W = \tfrac{1}{2}(c_{11}V_1^2 + 2c_{12}V_1V_2 + c_{22}V_2^2 + 2c_{13}V_1V_3 + \cdots) \tag{3}$$

2.8 ENERGY IN TERMS OF THE FIELD

The energy stored in an electrostatic system can be calculated by knowing the field D and E everywhere. This results because the charge distribution gives rise to a unique field.[1] The field or the charge distribution completely defines the stored energy.

In a parallel-plate capacitor, Section 1.15, the stored energy is $W = \frac{1}{2}CV^2 = \frac{1}{2}A\epsilon V^2/d$, where C is the capacitance value and V is the potential between the two conductors. The field E is V/d where d is the spacing between conductors. Substituting for V in the energy expression yields

$$W = \tfrac{1}{2}A\epsilon E^2 d \tag{1}$$

Recalling that $\epsilon E = D$ and the Ad is the volume of the field E and D, W can be written as

$$W = DEV \tag{2}$$

where D and E have the same direction.

In every field the region between closely spaced equipotentials can be treated as the equivalent of a capacitor (see Section 1.2). To find the total energy all such volumes must be considered. If the volume between adjacent potential shells dV stores an amount of energy dW, the total energy stored is given by the integral over-all volume elements or

$$W = \int_V D \cdot E \, dV \tag{3}$$

where this vector dot product insures that only the component of E in the same direction as the field D is considered.

2.9 WHY FIELD CONCEPTS ARE IMPORTANT

All electric energy is stored in fields. Conductors are used to define the geometry of this storage. At dc and low frequencies the largest percentage of system energy is stored in components. For example, the energy stored in a capacitor can be configured in a tightly wrapped cylindrical spiral. Components are usually selected based on frequency, and as the frequency of interest rises, the energy stored per unit of current or voltage decreases.

All conductor geometries have capacitance, and therefore these geometries allow energy to be stored in their surrounding space. The total

[1] A uniqueness theorem in electrostatics is proved in many advanced tests on static electricity. For example, see Smythe, "Static and Dynamic Electricity" 2d Ed., p. 25, McGraw-Hill, New York, 1950.

energy stored in a configuration of components thus rests both in the inner space of the components and in the space between interconnecting conductors. Obviously when the frequency of interest is high enough, the energy stored in the components disappears and the conductor geometry becomes a distributed component system.

If one keeps field concepts in mind, the transition from discrete circuits to high-frequency rf circuits is continuous; rf is no longer a separate phenomenon. This book attempts to show that an understanding of field concepts provides the best basis for understanding grounding and shielding problems. It also serves well to bridge the gap with rf phenomena.

3

Applying Electrostatics to Practical Processes

3.1 GENERAL

The electrostatics discussed in Chapters 1 and 2 provide the basis for much of the circuit behavior found in instrumentation. The bridge from electrostatic behavior to dynamic or ac phenomena is a short one. It is unfortunate that the concepts applied to electrostatics get pushed aside so easily once this step has been taken. Circuits are not always confined to interconnecting wires and to discrete elements, and the concepts of electrostatics explain much of this resulting phenomena.

Magnetic processes are not to be avoided. They are bypassed for the moment for convenience. This omission creates no loss at this time.

3.2 CURRENT IN CAPACITORS

The idea that a conductor or a group of conductors can hold a charge has been discussed. These charge distributions result in electric fields E and potentials V at points in space and on the conductors. Induced charges result in earthed or grounded conductors. The potential-charge relationships between elements of a system are defined by the geometry of the system. These ratios are the familiar capacitances and elastances of the system.

If the potential differences between elements are varied, the charges must adjust according to Eq. 2.4(1). The fact that the charges do vary implies a current flow. If an induced charge changes, then the current

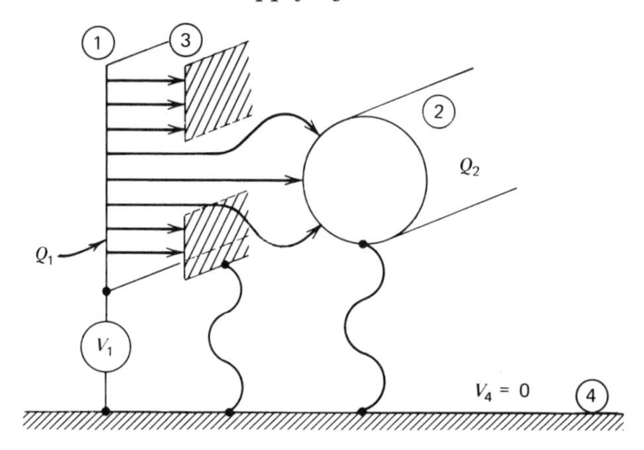

Figure 3.2 An example of self- and mutual capacitance.

flows in the earth plane. The charge in an element that is isolated or free from contact with other conductors cannot change. Current must flow in the wires or conductors forcing the change in potential.

Consider the self-capacitance c_{11} and the mutual capacitance c_{12} in Figure 3.2. If a mechanism V_1 exists for altering the charge Q_1 and if $V_2 = V_3 = 0$ (note that these conductors are connected to a zero potential), then

$$Q_1 = c_{11}V_1 \quad \text{and} \quad Q_2 = c_{12}V_1 \tag{1}$$

Differentiation of (1) yields

$$\frac{dQ_1}{dt} = c_{11}\frac{dV_1}{dt} \quad \text{and} \quad \frac{dQ_2}{dt} = c_{12}\frac{dV_1}{dt} \tag{2}$$

but the derivative of charge is current I, therefore

$$I_1 = c_{11}\frac{dV_1}{dt} \quad \text{and} \quad I_2 = c_{12}\frac{dV_1}{dt} \tag{3}$$

If V varies sinusoidally, that is, $V = V_m \sin \omega t$, then

$$I_1 = V_m\omega c_{11} \cos \omega T \qquad I_2 = V_m\omega c_{12} \cos wT \tag{4}$$

The ratio of voltage to current is reactance X_c:

$$X_{c_{11}} = \frac{1}{\omega c_{11}} \qquad X_{c_{12}} = \frac{1}{\omega c_{12}} \tag{5}$$

The flow of current I_1 or I_2 takes place in conductor ④ (the common zero potential) and in the voltage source V_1. The idea that a complete loop is required for current flow seems to be violated. This difficulty is explained by thinking of the changing electric field between the conductors as being the equivalent of current flow. This idea does provide the missing link that closes the current loop.[1]

In Figure 3.2 the change in induced charges Q_2 and Q_3 constitutes the reactive current flow. These charges reside at the zero potential of conductor ④, the earth plane.

3.3 VOLTAGE SOURCES

Many mechanisms exist that generate differences of potential. It is not the intent in this section to discuss the physics of potential generation but it is meaningful to mention a few typical sources. They include batteries, solar cells, thermocouples, or static potentials built up by friction. There are also potentials developed from changing magnetic-flux linkages, from the power mains, or from piezoelectric effects. However these potentials are generated, whether statically or time varying, they are carried on conductors to points of reference for application. These conductors are usually wires or cables, and wherever these conductors go an electric field containing electrostatic energy must also follow. If these potentials are time varying, currents flow in all the self- and mutual capacitances of the system. Figure 3.3 shows the electrostatic field around conductors attached to a 6-V battery.

The work required to move a unit charge from A to B over any path is 6 V. The battery furnishes this work when charge is moved through the battery. Since the static charges on conductors A and B are determined by the potential difference and the capacitance, no additional accumulation or modification of static charge is possible. If charge is moved it must flow in some load connected between the two conductors to avoid changing the charge on the conductors.

[1] Maxwell's equations require the term $\partial D/\partial t$ which is called the displacement current. See Ramo Whinnery, "Fields and Waves in Modern Radio," pp. 180–183. Wiley, New York, 1959.

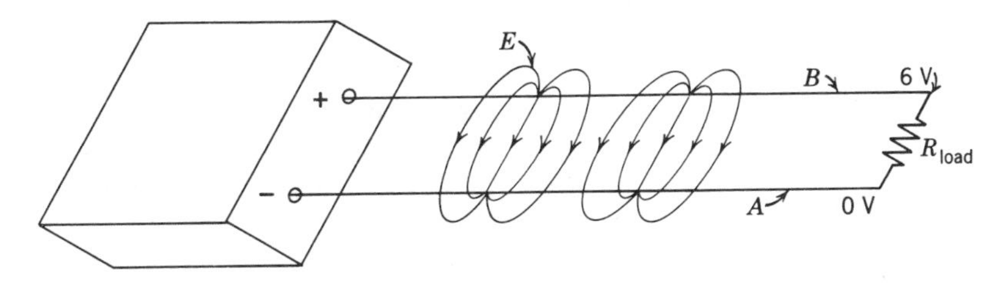

Figure 3.3a The electrostatic field about conductors and a battery.

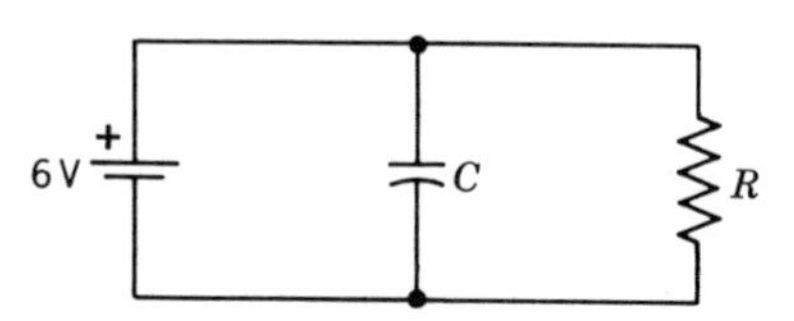

Figure 3.3b The circuit representation of Figure 3.3*a*.

When the structure in Figure 3.3*a* is examined in a circuit sense, as in Figure 3.3*b*, it is obvious that charge flows at a steady rate in the resistor R but remains static in the capacitor C. This circuit is very elementary but it is presented to show how closely tied electrostatic concepts are to circuit ideas.

3.4 ELECTROSTATIC SHIELDING

The word shield is commonplace in electronics. The idea is deceptively simple and yet is a source of much difficulty. The idea takes on the form of shielded wires, shield plates, shield boxes, metal screens, etc. Electrostatic shielding is just the property of Figure 2.5*b*. When conductor ② surrounds conductor ①, then the potential on ① cannot influence the charge on any other conductor. Stated another way, mutual-capacitance terms such as c_{13}, c_{14}, . . . , and so on, are zero. In practice these mutual-capacitance terms are not zero and they do permit unwanted effects. This problem will be treated in detail in Chapter 4. Conductor ② is called an electrostatic shield, or just a shield.

The mutual-capacitance terms between conductors within a shield to conductors outside of a shield are zero. If there are several conductors within the shield enclosure the mutual-capacitance coefficients of these conductors are not zero. The values of these capacitances will vary depending on the geometric relationships between these inner conductors and the shield conductor itself.

It is very commonplace to talk about grounding the shield. This is somewhat equivalent to the earthing of conductors required in Section 2.4. The intent there was to guarantee that all conductors were at zero potential so that the mutual-capacitance terms could be easily calculated. Grounding the shield is a more sophisticated problem and is treated fully in Section 4.4. It is correct to say that a shield can be at any potential and still provide shielding. This statement is true in the sense that relative changes in the conductor potentials within the shield have no influence on conductors outside of the shield. Also, changes in the potentials of the conductors outside of the shield have no effect on the

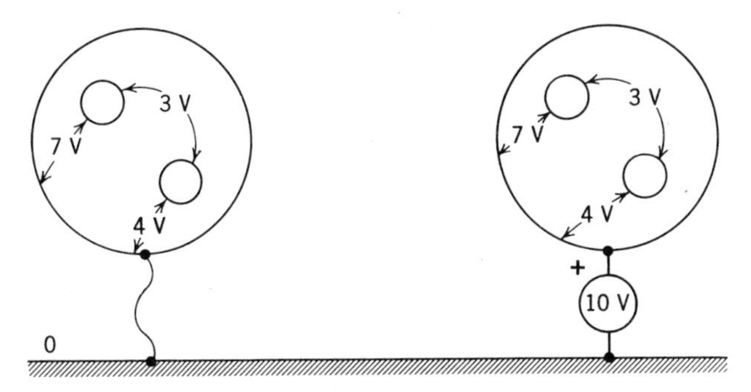

Figure 3.4 Shielding effect of a conductive enclosure.

relative potential of conductors within the shield. These statements do not require that the shield be earthed or defined in any way. The only requirement is that the conductors under discussion be fully surrounded by a conducting surface. This property is demonstrated in Figure 3.4. The potential differences for conductors within the shield in case (a) remain the same after the shield has been externally charged to 10 volts as in case (b). Note that the conductors within the shield are insulated. The internal potential differences exist only because of charges on these conductors. Unless these charges are altered, the potential differences remain the same.

3.5 THE EARTH PLANE

Every conductor has a finite resistance and the earth is no exception. The earth is used as a conductor in power systems and potential differences exist between points on the earth. For this reason one must be careful before considering the earth as a zero of potential. If one point is defined as zero potential, then the assumption that a nearby point is also at zero potential will usually be incorrect.

The electrical properties of the earth vary with the seasons of the year and with geographic location. The high-frequency or rf characteristics are not the same as power or low-frequency characteristics. Because of these unknowns the earth is usually not used as a signal-carrying conductor in instrumentation. Its influence cannot be avoided, however, as test structures, buildings, ac power, and so on, are all ohmically involved with the earth. These connections are unavoidable and result in some basic problems in instrumentation. Subsequent chapters deal with these earth ties and the best practices to minimize their effects. The earth as a conductor is discussed in Chapter 10.

Table 3.1 Table of Typical Capacitances

Description		Capacitance	
$\frac{1}{2}$-W carbon resistor, end to end		1.5	pF
2-cm-diameter sphere		1.11	pF
Parallel plates, $\frac{1}{4}''$ spacing/ft^2		114	pF
Two twisted No. 23 wires, HF insulation, per foot		25	pF
Two twisted No. 20 wires, cloth insulated, per foot		20	pF
Center wire to shield, RG 58, per foot		33	pF
Two-wire shielded conductors, No. 1 and No. 2 under a shield conductor No. 3. Capacitance per foot	c_{13}	64.9	pF
	c_{11}	77	pF
	c_{12}	43	pF
Outer shield capacitance to an earth plane (RG 58 on a tray) per foot		25	pF
Mutual capacitance from			
(a) Center wire of an RG 58 cable to a metal tray, per foot		0.15	pF
(b) Center wires of a two-conductor shielded cable to a metal tray, per foot		0.5	pF
Primary-to-secondary capacitance of a 20-W transformer		0.001	μF
Case-to-circuit of an HP200CD oscillator		610	pF
Soldering iron element to case (20-W)		40	pF
Man standing on an insulator to earth		700	pF
Relay coil to relay framework		50	pF
Pin-to-pin capacitance on an Amphenol connector		2	pF
Bonded strain-gage element to structure		140	pF
Thermocouple to structure		100	pF
Crystal transducer to case		30	pF
Capacitance of one No. 22 insulated wire in a bundle of No. 22 wires, per foot		40	pF
2N1192, element-to-case		2.5	pF
AB potentiometer, case-to-element		17	pF

3.6 TYPICAL CAPACITANCES

A group of representative capacitances are listed in Table 3.1. These values should be used with some caution as they are in many cases only approximate.

3.7 ROOM PICKUP

An electrostatic field at power frequencies exists in most inhabited areas. This field results from various forms of lighting, power distribution, zip cords, powered instruments, machinery, and so on. The field originates on open wiring and terminates on various grounds or conductors. In general, the field configuration is extremely complex. A person standing in a room adds to the complexity of the room's field. He assumes some ac potential other than the zero reference of a convenient piece of conduit unless he is touching that conduit. For this reason, a person is often described (improperly) as an antenna in that he picks up or couples power-frequency potentials.

A room can be thought of as a large capacitor. The lighting and conductors in the ceiling are one side of the capacitor and the floor or earth is the other side. A person in a room is literally standing in the middle of a capacitor.

The potentials in the middle of a typical room are difficult to measure. The presence of a probe or conductor for a measurement defines a new field with an equipotential along the path of the inserted conductor. With the field so modified, any measurement is invalid.

If a probe can be properly inserted to measure potential differences, the probe must not require charge or current to function as none is available in the free space. An instrument capable of measuring the field must therefore supply its own charge so that the field being measured is not modified by the process. It should be apparent that a plot of the electric field in a room would be almost impossible to make. This does not alter the fact that the field exists.

The induced reactive current flow per square foot of surface area in the earth plane of a typical room is about 100 nA at 60 Hz. This number can vary considerably and it should be used with caution. At power voltages this is an effective capacitance of about 2 pF.

4

Practical Shielding of Instruments

4.1 THE AMPLIFIER SHIELD

Consider an electronic device completely contained within a metal box. Further assume that the device is self-powered and no circuit conductors enter or leave the box. This circuit, as shown in Figure 4.1*a*, is completely shielded from external electrostatic influences (see section 3.4). The symbology indicates that a potential difference V_{13} between conductors ① and ③ is amplified to the value $-AV_{13}$ and this potential difference V_{23} appears between conductors ② and ③. Conductor ③ is called the zero signal reference conductor as it is common to V_{13} and V_{23}.

Mutual capacitances can be calculated by 2.4(1), but generally, for a typical circuit, would involve literally hundreds of conductors. The mutual capacitances that involve the shield enclosure are of present interest. The significant capacitances for an element of gain are shown

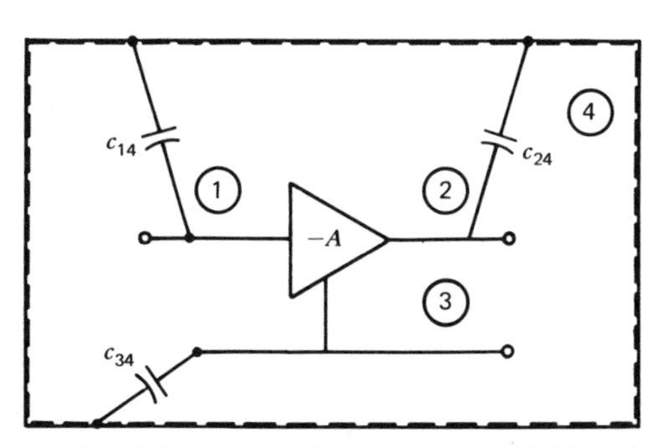

Figure 4.1*a* Mutual capacitances to the shield enclosure.

in Figure 4.1*a*. The effect of these capacitances upon the operation of the gain element becomes apparent in the equivalent circuit in Figure 4.1*b*. The mutual capacitances form a feedback structure around the gain element. These capacitances cannot be avoided but the obvious feedback process can be eliminated by ohmically tying the shield enclosure to conductor ③. This shorts out capacitance c_{34} leaving only capacitance c_{14} and c_{24}. This circuit is shown in Figure 4.1*c*. The feedback capacitances of Figure 4.1*b* are related to the geometry of the conductors. The placement of conductors could reduce this feedback but the preferred solution takes the form of Figure 4.1*c*.

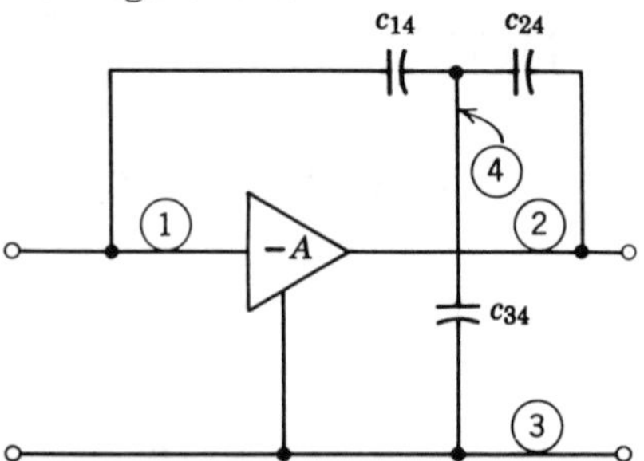

Figure 4.1*b* Mutual capacitances shown as circuit elements.

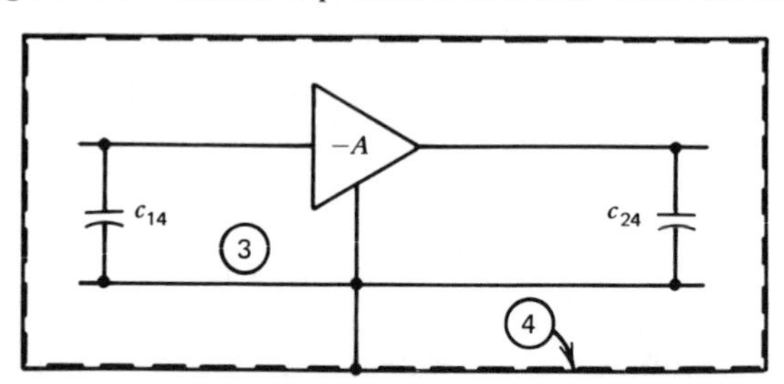

Figure 4.1c Elimination of undesirable feedback by eliminating c_{34}.

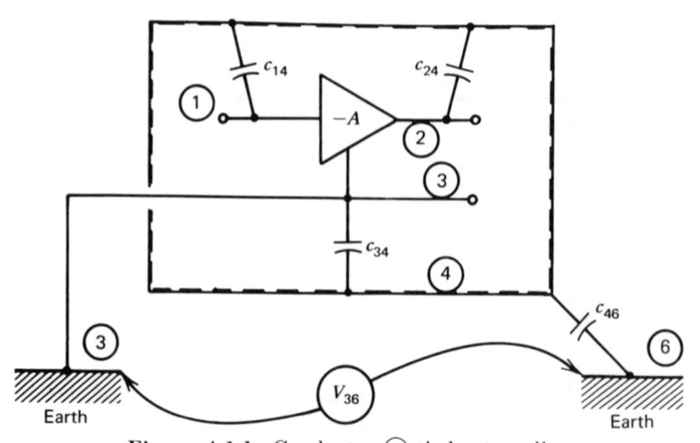

Figure 4.1*d* Conductor ③ tied externally.

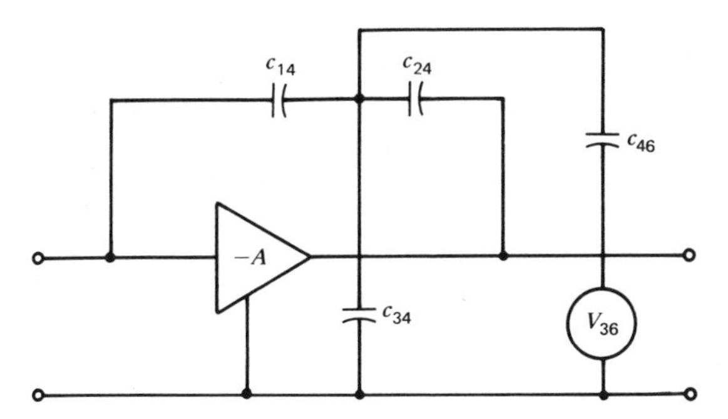

Figure 4.1*e* Circuit equivalent of Figure 4.1*d*.

If conductor ③ is taken outside of shield ④, a new phenomenon results. Figure 4.1*d* shows how an external mutual capacitance c_{46} now influences the circuit. If conductors ③ and ⑥ have a potential difference V_{36}, current results that flows in capacitances c_{34} and c_{46}. The path is ③→④→⑥→③. Since c_{34} is a feedback element, potential difference V_{36} is mixed with signals being processed by the gain element. This phenomenon is perhaps clearer in the equivalent circuit in Figure 4.1*e*. Again this influence is eliminated when c_{34} is shorted out.

The first rule of shielding results from the foregoing discussion.

Rule 1. *An electrostatic shield enclosure, to be effective, should be connected to the zero-signal reference potential of any circuitry contained within the shield.*

Conductor ③ exits the shield ④ in Figure 4.1*d*. This would appear to be acceptable provided Rule 1 above is followed. This conductor is then an extension of the shield and does not violate the shield-enclosure idea. Problems do result, however, and these are treated in the following sections.

4.2 SIGNAL ENTRANCES TO A SHIELD ENCLOSURE

The gain element in Figure 4.1*a* is impractical without input and output connections. Conductors that carry the signal to and from any amplifier are called signal conductors. For example, conductors ① and ③ are signal conductors in Figure 4.1*a*. Signal conductors are usually enclosed in a braided metallic sheath or shield, and this cable is called shielded wire If two conductors are within the shield it is called two-conductor shielded wire. This shielded wire is used to transport the signal from its source to the amplifier and can be thought of as an extension of the electrostatic enclosure of Figure 4.1*a*.

A shield enclosure is effective when Rule 1 is applied. This rule places no restriction on the shield potential relative to the external environment. This is the key to connecting signal conductors to a gain element. Since

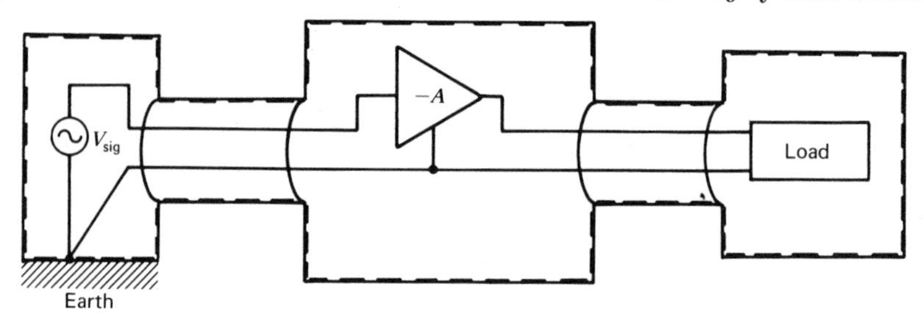

Figure 4.2 An extended shield enclosure including signal lines.

the shield must be at zero-signal reference potential, and since the signal is often derived from some reference point in the external environment, the shield is automatically defined at this external reference potential. There is no other choice.

Figure 4.2 shows a gain element and its shield enclosure. The input and output connections are two-wire shielded conductors. The input signal zero is ohmically connected to an earth point. When the shield is tied to this same earth potential Rule 1 is applied and the system is correct. This statement is very important:

Rule 1 requires that the shield must be tied to zero-signal reference potential. If the signal is earthed or grounded, the shield becomes earthed or grounded. Earthing or grounding the shield makes no sense if the signal is not earthed or grounded.

4.3 SHIELD CURRENTS

The electrostatic enclosures shown in Figure 4.2 often parallel several external conductors. For example, long runs of shielded wires are contained in raceways, in conduit, in floor wells, in parallel with other wires, in racks, or along floors. These neighboring conductors (grounds) are usually at differing potentials. In particular, these potentials are not the zero-signal reference potential of the shield enclosure.

These neighboring potentials cause currents to flow in the mutual capacitances between conductors. In Figure 4.3, current flows in loops such as ①→②→③→① or ①→③→④→⑤→①. This current flows in shield segments only and not in signal conductors. If this were to happen, unwanted, pickup would result.

4.4 SHIELD-DRAIN DIRECTION

Rule 1 requires that the shield be connected to zero-signal reference potential. No statement is included as to where this connection should

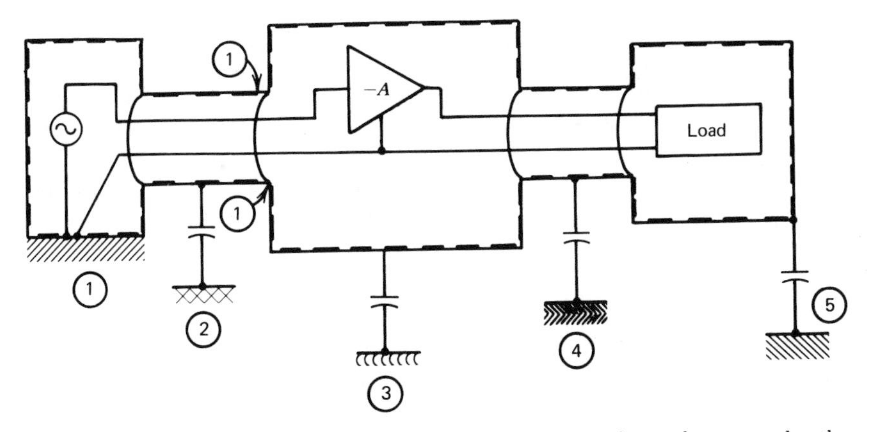

Figure 4.3 Mutual capacitances between an electrostatic enclosure and other earths and grounds.

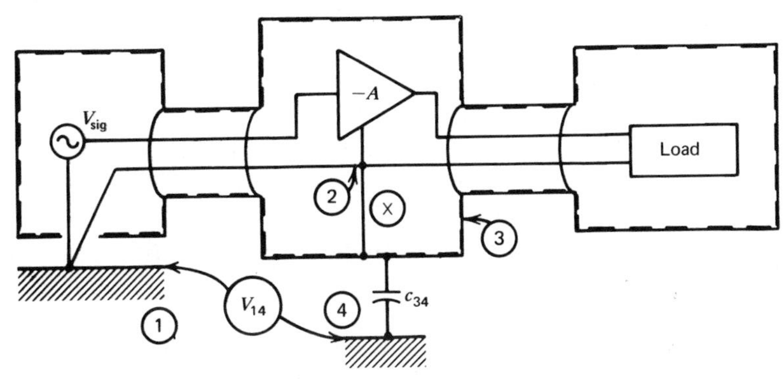

Figure 4.4 An incorrect tie between shield and the zero-signal reference potential.

be made. The connection is correctly made in Figure 4.2. An incorrect connection ⊗ is made in Figure 4.4 to illustrate the difficulty. The potential difference V_{14} causes current to flow in capacitance c_{34} in loop ①→②→③→④→①. If current flows in signal conductor ②, unwanted pickup results. To avoid this process, a second rule can be formulated.

Rule 2. *The shield conductor should be connected to the zero-signal reference potential at the signal-earth connection.*

This procedure ensures that parasitic currents will flow in the shield only and not flow in the signal conductors. The shield can be thought of as a drain path to carry unwanted current back to an earth point. Stated again:

Rule 2 requires that shields be connected so that shield currents drain to signal-earth connections.

The pickup that results from unwanted current flow in a signal conductor depends on cable length, frequency, and conductor size.[1] A possible problem occurs when the pickup magnitude approaches that of the signal being amplified in the frequency band of interest. Some systems have filters that electrically screen out these undesirable signals. This discussion is perfectly general and is intended to show the mechanics of pickup, not to argue whether the rules should or should not be followed.

4.5 SHIELD CONNECTIONS—SEGMENTS

By Rule 1, the electrostatic enclosure should be at zero-signal reference potential. If the shield is split in sections *Rule 2* places a constraint on the treatment of these segments. The rule requires that the shields be tied in tandem as one conductor and then connected to zero-signal reference potential at the signal-earth point. If the shield segments are individually treated the difficulties exhibited in Figure 4.4 can be expected.

Rule 2 is sometimes intentionally violated within an instrument amplifier. This problem is treated in Section 4.13. Shield connections that permit current to flow in an output or high-signal-level conductor are often ignored. The pickup here, as a percentage effect, is usually very low. Shield-drain processes in input conductors should be closely watched as the pickup here is subject to amplification. It is usually not too difficult to follow Rule 2 everywhere to avoid this and other difficulties that can result.

Rule 2 can be followed only when two-conductor shielded wire is used, as in Figure 4.2. Single-shielded wire (coax) obviously requires that the shield and zero-signal reference conductor be one and the same. Since unwanted current can only flow in the shield, pickup cannot be easily controlled with this type of cable. Fewer problems result when output cables are coaxial but new problems such as cross talk can arise. These effects are discussed in subsequent chapters. Also see the notes on shielding charge amplifiers in sections 6.3 and 6.4.

[1] Reactive currents of up to 1 mA often flow in circuits such as Figure 4.1*d*. If conductor ③ is 10 ft of No. 25 wire, its resistance is 0.32 Ω and the pickup is 0.32 mV rms. At the output of an amplifier with gain of 1000 this is 0.9 V peak-to-peak, a sizable signal indeed.

Even when the pickup is outside the band of interest, filtering may be ineffective. This results when nonlinear effects such as rectification occur and when the filtering is placed after the point of nonlinearity. Since the nature of the pickup cannot always be predicted, it appears worthwhile to take heed and follow good instrumentation practice.

4.6 POWER ENTRANCES

The instruments described in previous sections require operating power. Utility power is preferred over alternate sources such as batteries or solar cells. These alternate sources contain no 60-Hz influences but they are generally impractical. The use of utility power poses problems but the resulting ac effects can be reduced to acceptable levels.

If transformer action is employed, power can enter a shield enclosure without violating Rule 1. In practical power transformers, the materials used for shielding are usually copper or aluminum. These materials are nonmagnetic and they do not hinder a magnetic field. The magnetic field effectively crosses the shield boundary and couples energy from the primary to the secondary of the transformer.

4.7 POWER-TRANSFORMER CONVENTIONS

It is the custom to draw a power transformer with the core shown as parallel lines between the primary and secondary coils. Since all of the transformers used in the following sections use magnetic core material, and since these core representations add nothing to the understanding, these lines will be deleted from all drawings. Effective transformer action requires that the primary and secondary coils be properly coupled magnetically. One practice that is often used places the primary coil physically over or under the secondary coil. The nonmagnetic shield used to separate the primary from the secondary coil must take on a complex shape. A simple transformer representation illustrates none of this complexity.

In practice, a transformer shield is a wrap of copper or aluminum over a coil or over several coils. It constitutes a single open turn as it usually threads the core along with the coil. As a turn it can not close on itself as this would constitute a shorted turn. A transformer with a shorted turn is considered defective.

Shields can be of differing quality. They can be simple and only fill the space between coils or they can be extended to cover the sides of the coils as well as the leads exiting the coils. They can be made electrically "watertight" or full of "leaks." In the following discussions the shields are assumed perfect to illustrate their use and effectiveness. See Section 8.4 for further treatment of shielded transformers.

4.8 POWER TRANSFORMER WITH A SINGLE SHIELD

Figure 4.8 shows a typical power-transformer entrance into a shield enclosure. The transformer has one shield ④. This shield is one segment

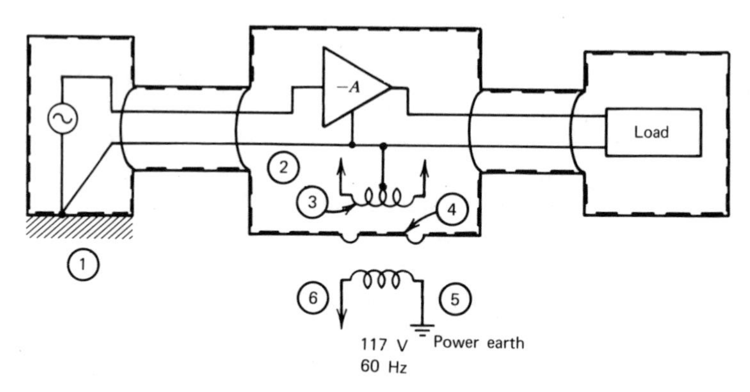

Figure 4.8 A power-transformer entrance into a shield enclosure.

of the total shield and is shown connected by Rule 2 to the remaining shield enclosure. The transformer shield is usually close to both the primary and secondary coils of the transformer. These coils have power-frequency voltages on their turns. Each turn has a capacitance to the shield and this potential difference causes current to flow in the coil-to-shield capacitance. The lumped effect of the coil-shield relationship is that of a single capacitance in series with a fraction of the winding voltage. This effect takes place on both sides of the shield and causes currents to flow in two separate paths.

4.9 COIL-TO-SHIELD CAPACITANCE

Consider a coil and shield arrangement as shown in Figure 4.9*a*. This coil is a single layer of conductors, each loop having a capacitance c_m to the shield. If the shield and one end of the coil are at zero potential, the current flowing in each capacitance will depend on the voltage level at each turn. Assume an n-turn coil with a potential V across the coil. The potential on the mth turn is mV/n. The current in the mth capacitance is $Vm/n \cdot 1/X_c$. The total current in all n conductors where n is large is

$$I = \sum_{m=1}^{n} \frac{Vm}{n} \frac{1}{X_c} = \frac{V_n}{2X_c} \tag{1}$$

The capacitance of the conductors to the shield is the parallel combination of all turns capacitances or nc_m. The current of (1) flows as if this capacitance is connected to the midpoint on the coil as shown in Figure 4.9*b*.

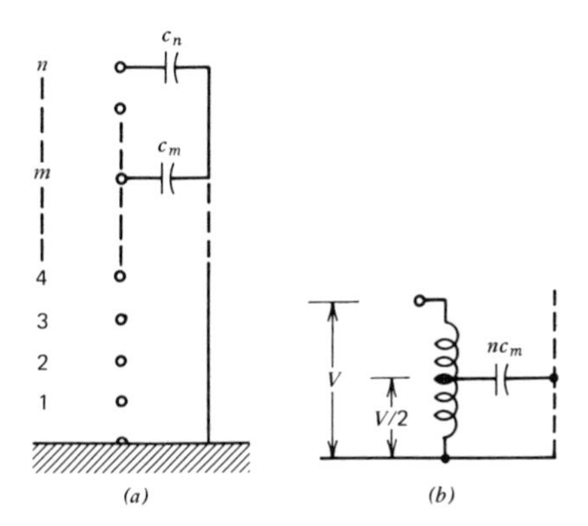

Figure 4.9a A transformer shield and coil.
Figure 4.9b Effective shield-to-coil capacitance.

Transformer coils are usually layer-wound. The last layer of turns will be next to any shield placed over the coil. If the center tap of the coil is pulled out from an inner layer, then the shield will in general not be physically near this point. The turns in an outer layer screen the turns from all internal layers (low mutual capacitance). This means that the only turns involved in coil-to-shield reactive current flow are in the outer layer. Since the full coil has many layers, the potential difference across this last layer is usually only a fraction of the total coil voltage. For example, if 10 layers are involved in a 120-V coil, then 12 V appears across each layer. The potential difference between the midpoint and the ends of the outer layer is 6 V. If the start of the coil is at 0 V, the outer layer midpoint is at 114 V. If the start of the coil is at 120 V then the outer layer midpoint is at 6 V.

Current flow to the shield from this outer layer is described by Eq. 4.9(1). The capacitance is essentially that of the last layer of coil to the shield or nc_m. The potential differences V_n must be measured between the midpoint of the outer layer and the shield. If the shield is at zero potential and the coil end adjacent to the shield is connected to 120 V, then the potential difference forcing shield current is 114 V.

Utility power is usually furnished with one conductor at or near earth potential. The coil-shield arrangement shown in Figures 4.9*a* and 4.9*b* is thus typical of the transformer primary-to-shield conditions encountered in instrumentation.

The center-tapped secondary coil in Figure 4.8 is also typical. The preceding arguments show that the shield is rarely balanced with respect to this center tap. On the primary side the potential V_n depends on the sense of the power connection. This is why instruments with unshielded or one-shielded transformers often operate best with a preferred power-cord orientation. This discussion should make it obvious to the reader that transformer diagrams are quite misleading; the diagrams would imply symmetry whereas in reality there is unbalance.

4.10 THE SINGLE TRANSFORMER SHIELD AND ITS CONNECTIONS

The coils in the transformer of Figure 4.8 act as simple voltage sources in series with coil-to-shield capacitances. These potentials circulate currents in the shield structure. Figure 4.10*a* shows these sources and the effective series capacitances.

Assume that there is an ohmic path between the earth ① and the earth ⑤ and also that these conductors are at the same potential. (If they are not at the same potential, the difference can be absorbed by adding it to the potential at ⑥.)

The current loop in the primary is ①→⑦→⑥→⑤→①. This current path is outside of the shield enclosure. The current loop for the secondary, however, is ①→②→③→④→⑦→① and this current flows in the signal-input conductor ② and must be considered. If c_{34} has a 1 MΩ reactance at 60 Hz, and the transformer potential difference from ② to ③ is 5 V, then 5 μA will flow. If the conductor ② has 2Ω of resistance, 10 μV of pickup will result. In many instrumentation processes this is excessive.

The transformer shield can be alternately connected to the zero-signal reference potential at the gain element. This confines the secondary circulating currents but extends the primary-coil current flow. Figure 4.10*b* demonstrates the new problem. The secondary current flow in capacitance c_{34} follows the loop ④→②→③→④. This loop does not involve the resistance in conductor ②. The primary current flow in capacitance c_{46} follows the loop ⑥→④→②→①→⑤→⑥ and this current does flow in the length of conductor ②. The calculation for the circuit in Figure 4.10*a* showed a 10 μV pickup. In Figure 4.10*b* the same circuit values result in 200 μV pickup. (Note: the capacitance c_{46} has 100 V as a

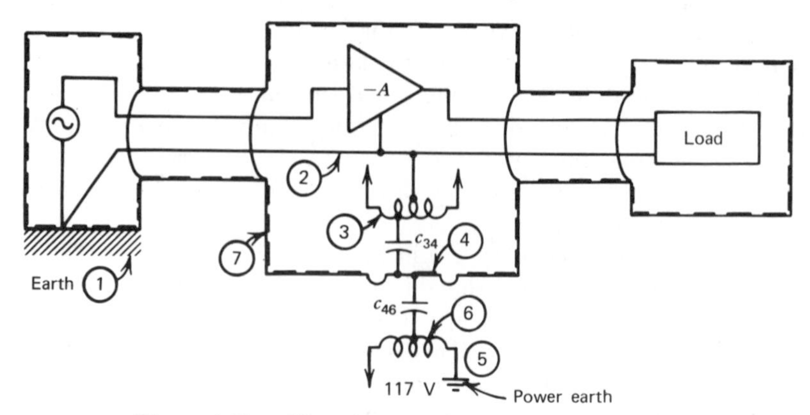

Figure 4.10*a* The primary and secondary capacitances.

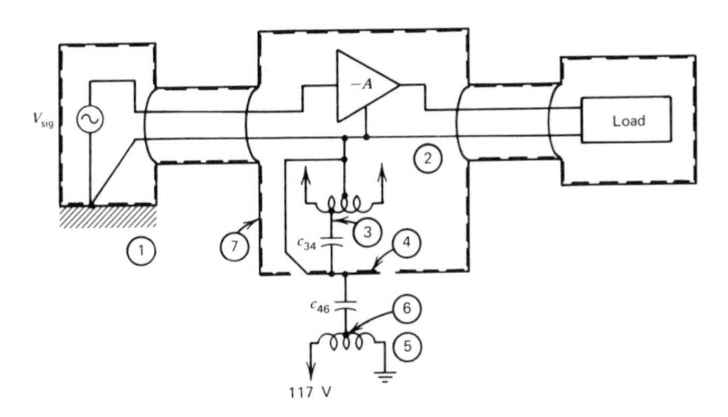

Figure 4.10b Transformer and segmented shield.

source potential, nearly the entire primary voltage.) This pickup is amplified by the gain element.

The shield connection shown in Figure 4.10*b* is the most commonly used. The difficulties it creates for the application shown in Figure 4.10*b* are obvious. If the external earth-signal connection is made at the output or signal-termination point, the difficulty is not as severe. Currents still flow in conductor ②, but in the output side where the pickup is not amplified by the gain element.

4.11 THE DOUBLE ELECTROSTATIC SHIELD

The current that flows in the zero-signal reference conductor in a single-shielded transformer cannot be eliminated. No placement of the transformer shield avoids the problem. In some applications the pickup effects can be minimal and a single shield is adequate.

To contain the currents caused by the transformer voltages, a second shield must be added. This shield is placed directly over the first shield but insulated from it. The shield nearest the primary coil is called the primary shield and similarly the shield nearest the secondary is called the secondary shield.

The proper connections for these two shields are shown in Figure 4.11. The secondary shield is segmented and connected to zero-signal reference potential at the gain element. The primary shield is often connected to the rack or frame of the supporting console. The primary-shield currents flow in the path ⑤→⑥→⑦→⑧→⑤ through capacitance c_{56}. The secondary-shield currents flow in the path ②→③→④→② through capacitance c_{34}. In both of these loops, the current does not flow along the zero-signal reference conductor ②.

A potential difference usually exists between conductors ⑦ and ⑧ in

Figure 4.11. This potential adds to the current flow and is equivalent to a modified tap point on the coil for the capacitance c_{56}. The potential at the primary connection ⑦ is determined by the utility company and corresponds to their power neutral.

The primary shield (conductor 8) in Figure 4.11 lies between the primary coil and the secondary shield, and serves its function by intercepting the electric flux from the primary coil. An unshielded power cord followed by a shielded segment of conduit usually follows this along the power link. The primary shielding is thus segmented, partially complete, and connected to various and unknown potentials referred to indiscriminately as ground. After the shield serves its purpose in the transformer, the full enclosure of the power distribution is not maintained. Fortunately it is not important as this procedure would be difficult indeed. For this reason the primary coil shield is not shown as a full enclosure in all diagrams.

The connection point for the primary shield in Figure 4.11 should be considered. The potential difference between earth connections ① and ⑧ will cause current to flow in capacitance c_{45}. This capacitance in a 10-W transformer may be 0.001μF. The loop path is ①→②→④→⑤→⑧→① and this includes the zero-signal reference conductor. If a 10-V potential difference exists, a 3-μA current flow results. In a 2-Ω line this is 6 μV of pickup. The proper potential for the primary shield is thus zero-signal reference potential. A connection directly to the gain-element zero-reference conductor negates the secondary shield. Therefore, the primary-

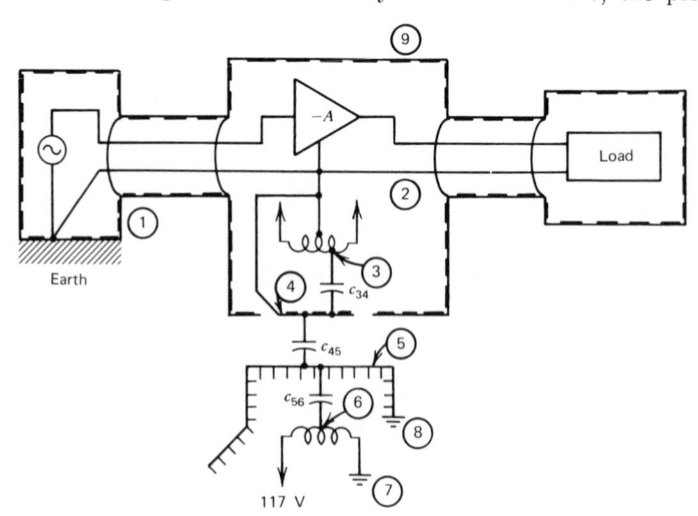

Figure 4.11 A double-shielded transformer.

shield connection must be made via another path to be effective. See section 4.15 for further discussion.

The transformer shields as shown in Figure 4.11 need only be moderately good to be effective. A hundredfold reduction in effects results if the conductors within the shield enclosure have mutual capacitances to the total external environment of 5 pF or less. The leakage or mutual capacitance from the primary coil to the secondary shield is c_{46}. The leakage or mutual capacitance from the secondary to the primary shield is c_{35}. These capacitances are not drawn out as circuit elements in Figure 4.11, but they are real and do exist.

The doubly shielded transformer provides some immunity from unwanted pickup caused by reactive current flow. It should be apparent that it is not the complete solution as the effects just described may be difficult to overcome. The differential amplifiers discussed in Chapter 5 are best suited for solving this and many other problems encountered in instrumentation.

4.12 SINGLE-ENDED AMPLIFIERS

The gain element in Figure 4.11 is called a single-ended amplifier. The distinguishing feature is the through connection of the zero-signal reference conductor. This through connection can be effective as a signal-carrying element as long as unwanted current does not flow along its length. The effects of the power transformer were discussed in Sections 4.10 and 4.11. The external connection to earth were discussed in Section 4.1. If two such external connections were made large currents could flow in this conductor. If this current flows in an input segment, the pickup is amplified. Current flow in an output segment may not be as damaging, but this is a relative matter. Therefore only one point along the zero-signal reference conductor should be earthed.

A philosophy accepted in some quarters is to short out offending potential differences so that two external connections can be made to the same zero-reference conductor. Even if the potential difference can be reduced to 0.01 V, the reference conductor must straddle this 0.01-V difference. Therefore some fraction of this 0.01 V will appear as an input signal. This solution is not suggested as practical or useful but is often used as a last resort when other possibilities are exhausted.

Two external ground connections to a zero-reference conductor cause a ground loop. Current will flow in any such loop and the nature of the current will depend on many unknowns. In other words, the current level may be low enough today, but tomorrow may be another story.

4.13 SEGMENTING THE AMPLIFIER SHIELD

All efforts in shielding are directed towards reducing unwanted pickup. The procedures outlined in Section 4.11 have a practical flaw. The shield structure ⑨ in Figure 4.11 is usually in close proximity to sensitive gain elements within the amplifier. This shield is connected to *one point only* and this is at the signal-earth connection by Rules 1 and 2. The distance to this connection can vary from a few feet to thousands of feet. At large distances the inductive effects and transmission-line effects will allow potential differences to be developed along the shield length. See Section 4.3. If a shield has potential differences along its length, it can be at zero-signal reference potential at one point only. These shield signals can thus couple into sensitive circuitry contained within the shield.

Amplifiers are usually manufactured with the local shield segmented and connected to the zero-signal reference potential at the amplifier. This avoids the direct contaminating effects indicated above. If this were not done amplifiers would rarely meet their noise specifications. There is no loss caused by this practice as the signal brought to the amplifier by the input shield is not significantly influenced by the amplifier-shield segment. The zero-signal reference potential the amplifier sees at its input terminals cannot be improved upon. The shield may be affected by its external environment but only a fraction of this contamination couples into the signal. It therefore makes good sense to shield the amplifier with the best available measure of the zero-signal reference potential. This practice is shown in Figure 4.13.

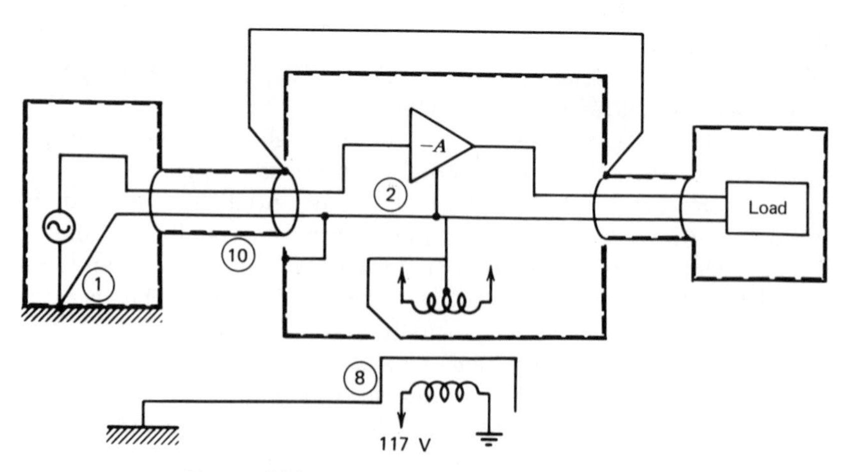

Figure 4.13 A practical double-shield system.

4.14 A SHIELD-ENCLOSURE RULE

The discussion in the preceding section on double shielding in power transformers lends insight into an important generalization. The first shield was used to surround all of the signal processes and this included signal conductors, the gain element, and the secondary of the power transformer. The second shield was used to enclose the power source insofar as it was practical and meaningful.

A third shielding rule would not be apparent from the above discussion alone. It is stated here and will be required to extend the application of the electrostatic shield in the next chapters. As the rule is applied its value will become more apparent.

Rule 3. *The number of separate shields required in a system is equal to the number of independent signals being processed plus one for each power entrance.*

The three shielding rules described in this chapter provide a basis for design. It may not always be necessary to apply them in their strictest sense; however, it is important to understand why they have been formulated. If a rule is not followed it is good design practice to understand the side effects and to calculate their magnitude if possible.

4.15 PRIMARY-SHIELD TIES

The primary shield ⑧ in Figure 4.11 should be connected to a point of zero-signal reference potential (see section 4.11). This can be accommodated by a separate conductor connecting ⑧ to ①. It is practical in some cases to use the shield conductor ⑩ for this connection. If a low-level signal is being processed over this long line, pickup may result because the shield current establishes a potential gradient along the shield's length.

If several signals are being amplified and they originate in the same zero-signal reference environment, one conductor can be brought back for connection to the several primary shields involved. A separate wire for each instrument is not required. If a shield is used for this connection it should preferably be the shield of a high-level signal. If this is not available, the separate conductor is preferred.

A subtle advantage is gained by returning the primary shield to earth via the input shield. This approach reduces the loop area ①→②→⑧→①, which includes the input signal conductor ②. This loop area subjects the input to possible unwanted magnetic pickup and the proper shield return reduces this possibility.

4.16 A NOTE ON LOCATING CURRENT LOOPS

Every circuit or system has a large number of possible current loops. Some of these loops permit unwanted signal pickup. The author is frequently asked, why did you pick that loop—what is the matter with this one? Often the loop suggested is an impossible one because current would have to return to zero potential and still traverse a further impedance to make the full circle. Admittedly the circuits are not conveniently drawn out, but the circuits are simple once they have been located. The author has found it convenient to use the following rules to locate current loops involving transformer coils:

1. Start with each zero-signal reference conductor.
2. For each transformer coil referenced to this potential, locate the possible mutual capacitances (leakage capacitances) that will permit current flow.
3. From the terminating end of this mutual capacitance locate all return paths to the zero-signal reference conductor.
4. Once a current path returns to zero-signal reference it cannot proceed further.
5. Consider all ground or earth points as tied together.

The following statements will help in locating current loops involving ground potential differences:

1. Start with each zero-signal reference shield.
2. Consider the mutual capacitance to any adjacent shield.
3. If this shield is at a potential other than zero-signal reference, then current can flow in the mutual capacitance.
4. From the terminating end of this mutual capacitance locate all possible return paths to the reference shield.
5. Once a current path returns to the reference shield it cannot proceed further.
6. Consider all ground or earth points as tied together.

5

The Differential Amplifier

5.1 GENERAL

The words *differential amplifier* have a very general meaning. There are many instrumentation problems requiring the use of differential amplifiers but not all instruments with this generic title will prove to be useful.

Many oscilloscopes provide a differential type of input circuit. Here, a common terminal plus two input terminals, often marked A and B, provide a signal entrance. The oscilloscope pattern demonstrates the subtraction of signals applied to terminals A and B; that is, the pattern is proportional to signals $A - B$. If $A = B$ the pattern remains unaffected. The value of signal difference, $A - B$, is termed the differential signal. The average value of signal $(A + B)/2$, which is applied to both inputs, is ignored when the oscilloscope is properly balanced.

An example of this operation may be useful. Figure 5.1 shows several

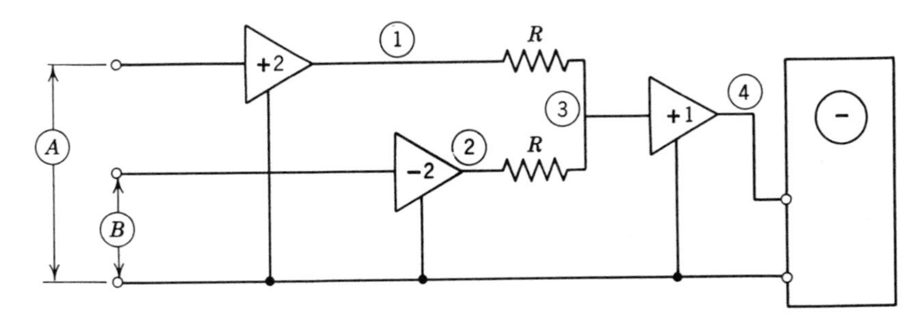

Figure 5.1 A simple differential amplifier.

Table 5.1 Performance of the Differential Amplifier in Figure 5.1

Input Voltages		Circuit Point Voltages			Output Voltages		Average Input Voltages
A	B	1	2	3	4	$A - B$	$(A + B)/2$
3	−3	6	6	6	6	6	0
0	4	0	−8	−4	−4	−4	2
4	0	8	0	4	4	4	2
10	4	20	−8	6	6	6	7
−10	−4	−20	8	−6	−6	−6	−7
7	1	14	−2	6	6	6	4
6	6	12	−12	0	0	0	6

gain elements connected to provide differential operation. The numbers within the triangles indicate the voltage gain provided by that segment. A table of signal levels for points within the circuit of Figure 5.1 are given in Table 5.1.

Note that the output ④ measures $A - B$ only and does not relate to the average value $(A + B)/2$. Signal $A - B$ is called the difference or differential input signal.

5.2 A BASIC INSTRUMENTATION PROBLEM

To place the differential amplifier requirement into better perspective, a problem basic to instrumentation is next treated. Consider a signal potential that is developed in one zero-signal reference region, and must

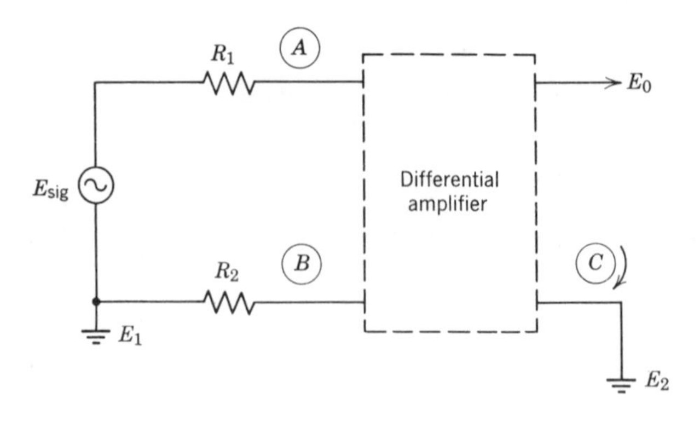

Figure 5.2 Two zero-signal reference regions.

be observed in a second zero-signal reference region. Figure 5.2 illustrates this condition. The need for a differential amplifier in the conventional sense becomes apparent if the zero-signal reference for the output signal is taken as the zero reference potential for observing the input signal ($E_2 = 0$). If we can assume that zero current flows in R_1 or R_2, then the potential at Ⓑ as measured from Ⓒ is E_1. This is just the zero reference potential of the signal measured from Ⓒ. The potential at Ⓐ is $E_1 + E_{\text{sig}}$. The potential difference E_{sig} can be measured from Ⓒ by subtracting Ⓑ from Ⓐ; that is, $E_{\text{sig}} = A - B$. The average value of the input $(A + B)/2$ is $(E_{\text{sig}} + E_1 + E_1)/2 = \frac{1}{2}E_{\text{sig}} + E_1$. Since the unwanted term E_1 is present, the amplifier in Figure 5.2 must not amplify this average value. In Figure 5.1 this exact problem was treated by using a differential amplifier.

The circuits of Figures 5.1 and 5.2 differ in application. The node containing potential E_1 is not available in Figure 5.1. An equivalent point could be constructed by juggling the composition of potentials Ⓐ and Ⓑ but this is academic. The circuit in Figure 5.1 assumed one zero-signal reference potential. The instrumentation problem in Figure 5.2 requires two such zero-signal potentials. Both circuits require a subtractive process (differential amplification) but, as will be pointed out, the demands on the circuitry in Figure 5.2 are rather significant. These demands can only be met by understanding the basic shielding and grounding processes involved.

5.3 INSTRUMENTATION DIFFERENTIAL AMPLIFIERS

The general instrumentation problem becomes clearer if the current flow in R_1 and R_2 in Figure 5.2 is considered. The potential difference $E_2 - E_1$ will cause current to flow in R_1 or R_2 if an electrical path exists between Ⓐ or Ⓑ and Ⓒ. The impedance levels involved vary between applications but a good measure can be obtained by considering a typical problem.

If the input signal levels are 10 mV and 0.1% accuracy is required, and if $R_1 = 1000\ \Omega$ and $R_2 = 0$, the current flowing in R_1 must be less than 10 nA. If $E_2 - E_1$ is 10 V, the impedance permitting current flow must be greater than 1000 $M\Omega$. If $R_2 = 1000\ \Omega$ and $R_1 = 0$, the problem is exactly the same. This point is very important. The 1000 MΩ figure was derived by considering signal levels, source, and termination characteristics only. It is a problem basic to transferring signals from one zero-signal reference potential to another. This argument alone supports the statement that many devices may be differential in character but not all are applicable in solving the basic instrumentation problem. It also points up the important fact that all conductive paths, reactive and resistive, must be considered, the amplifier being only one area of concern.

The resistances R_1 and R_2 in Figure 5.2 may be a part of a signal transducer or they may represent the conductor resistance of an input cable. Currents flowing in resistances R_1 and R_2, due to the potential difference $E_2 - E_1$, add to the signals at A and B. If resistors R_1 and R_2 are equal to 1000 Ω, 1 μA of current in both resistors would add 1 mV of potential to both input conductors referenced to the zero of input potential. This added 1 mV of potential is equivalent to changing the value of E_1 by 1 mV. We conclude that equal currents flowing in equal input resistances cause a new average input signal and this signal is not amplified by the differential amplifier.

Currents flowing in unequal resistances R_1 and R_2 can cause a differential or difference input signal. This was illustrated above when R_2 was 0 and R_1 was 1000 Ω. Since this difference signal is caused by unwanted current flow, instrument manufacturers specify instrument performance by referring to the source or *line unbalance*. This is simply the difference of resistance value $R_2 - R_1$. The larger this line unbalance the greater the unwanted differential pickup will be.

The resistances R_1 and R_2 that may be used are restricted in value by the instrument manufacturer. Instrument performance may be quite poor if these resistances are too high in value even if their difference value is small. This topic is also discussed in Section 5.15.

5.4 COMMON-MODE VOLTAGE

The average input signal for the differential amplifiers in Figures 5.1 and 5.2 is $(A + B)/2$. For Figure 5.2 this is $\frac{1}{2}E_{sig} + E_1$. This average common input signal is referred to as a common-mode signal. It is apparent that the average value can contain both wanted and unwanted signals and this can be confusing.

The testing and specifying of instruments with regard to rejecting unwanted signals are usually made with the understanding that the normal signal is kept at zero. If this is true, the average input signal is all unwanted in nature and the rejection measurement is straightforward.

The instrumentation problem in Figure 5.2 has an average input referenced to the output signal zero of $\frac{1}{2}E_{sig} + E_1$. If E_{sig} is zero, the common-mode voltage is just E_1. This leads to a very important result. Assuming zero input signal, the common-mode voltage and the ground differences of potential between zero signal reference conductors are one and the same thing. Unwanted signals are often orders of magnitude greater in level than normal signals. It is therefore accepted practice to say that *the common-mode signal is simply the ground-difference of potential between zero-signal reference points.*

Unwanted and undefined potential differences exist between the input zero-signal reference point and other grounds, such as power neutral. These potential differences can contaminate the signal; however, they are not, strictly speaking, common-mode voltages. If it is necessary to refer to these contaminating influences, the exact nature of the expected problem must be stated. Without a well-defined specification, the chances of communication are slight indeed and the expected difficulty may never receive consideration.

A second type of common-mode signal is often encountered in practice. In strain-gage applications, for example, the average potential of the input lines with respect to the input zero-signal reference conductor may be as great as 5 to 10 V. This average signal is usually one-half the bridge excitation voltage if one side of the excitation is assumed to be grounded. This excitation potential is common to both input signal lines and is therefore a common-mode signal. Since this signal is static it can cause an offset voltage if not rejected by the differential amplifier. Ground potential differences are usually at power frequencies or their harmonics and generally add ac content to the output signal if not rejected by the differential amplifier.

Common-mode rejection specifications should properly reference both types of signals as described above. It should be further noted that these common-mode signals may be interactive; that is, the rejection of one signal may depend on the absence of the other. This specification problem is not properly a part of any discussion of grounding but it is a difficulty frequently encountered. Because the two types of signal are often additive at the point of high impedance, care should be taken that the total common-mode signal is within the specified limits.

5.5 COMMON-MODE CONTENT

The potential differences that make up the common-mode signals encountered in instrumentation can vary in level from a few millivolts to thousands of volts and can vary in frequency from dc to rf.

Common-mode signals often occur at 60 Hz and harmonics of 60 Hz. In installations where the earth is a very poor conductor or in areas where several utility services are provided, earth differences of potential can exceed 50 V. Voltages of this magnitude are not often found, although pulse or transient common-mode signals of this level are rather frequent.

In systems where high-frequency power generation is involved, the common-mode signals can be composites of 60 Hz and this higher power frequency. Reactive effects will usually increase the level of common-mode difficulty at those higher frequencies. This results because leakage capacitances permit larger reactive currents to flow, and shield or earth inductances permit greater potential drops, which increase coupling effects.

Common-mode signals can be rf in nature. For example, radio broadcast energy can propagate along signal cables and can result in unwanted signal pickup. On occasions this rf energy can overload an input element creating noises or disturbances that are nonlinear in nature. However the rf interference manifests itself, it can prove to be a serious instrumentation problem. Further discussion about this topic is contained in Chapter 9.

5.6 COMMON-MODE REJECTION RATIO OR CMR

The ratio between the common-mode signal and the unwanted pickup at the input to an instrument amplifier is termed the common-mode rejection ratio. In the previous example a $10\mu V$ of pickup caused by a 10V common-mode signal represents a rejection ratio of 10^6. This is often expressed in decibels as 120 dB. Since the effects of pickup are measured at an amplifier's output terminals, the signal at the input is simply the pickup at the output divided by the gain setting. Stated as an equation, common-mode rejection (CMR) is simply

$$\text{CMR} = \frac{E_{\text{CM}}G}{E_{\text{out}}}$$

where E_{CM} is common-mode voltage, E_{out} is the output common-mode signal and G is the amplifier gain.

5.7 SOLUTIONS TO THE DIFFERENTIAL-AMPLIFIER PROBLEM

The differential amplifier shown in Figure 5.2 takes on many forms depending on specifications, size, component technology, and so on. The nature of these instruments and how they relate to the system via cables, source impedances, shielding needs, power entrances, and so on, are of importance to anyone involved in instrumentation processes.

The three rules developed in Chapter 4 can be used to arrive at an understanding of the instrumentation techniques that can meet the requirements outlined in Section 5.2. The first rule requires two shield enclosures. Rule 2 requires that these shields be connected to zero-signal reference potential at the signal-earth connections. Figure 5.7*a* shows these rules applied to our problem.

The gain elements marked A_1 and A_2 can vary in character over a wide range. The electronics can include modulators, choppers, amplifiers,

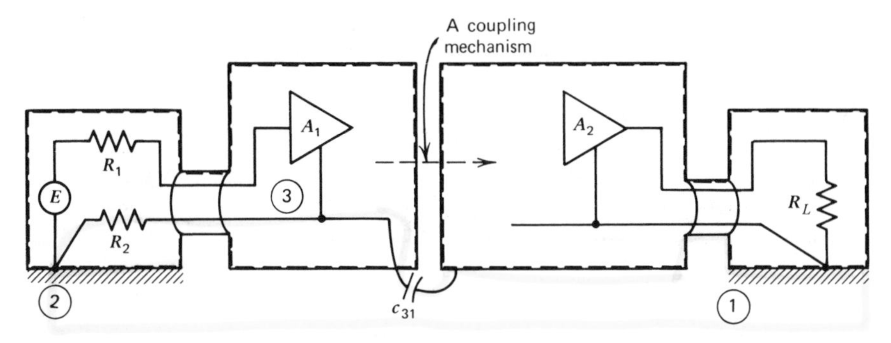

Figure 5.7*a* The two-shield enclosures in a differential amplifier.

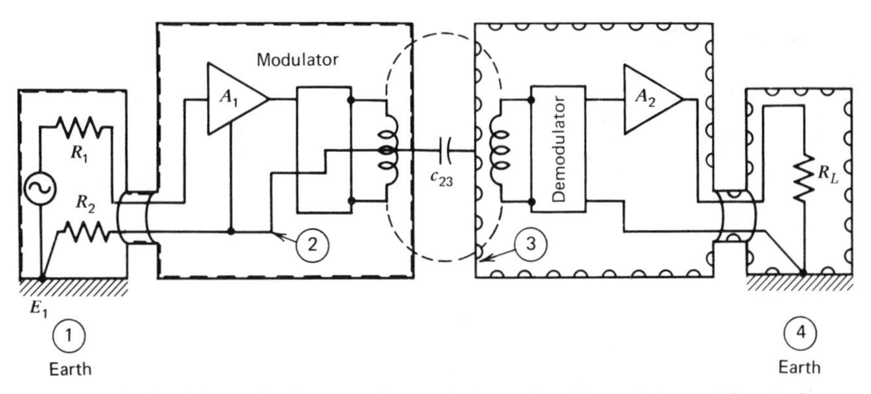

Figure 5.7*b* Magnetic flux coupling techniques in differential amplifier design.

transformers, or just passive elements. The gains can be distributed between A_1 and A_2 in any fashion depending on the designer's requirements. The obvious problem the designer faces is one of signal transfer from A_1 to A_2 without violating the shield enclosure. Note that a mutual capacitance term such as c_{31} permits current to flow in the path ②→③→①→② *and this current flows in* R_2. By the arguments given in Section 5.3, this reactance should exceed 1000 MΩ at the common-mode frequency. At 60 Hz this is approximately 2 pF.

The transfer of signal data from A_1 to A_2 in Figure 5.7*a* can be made by such techniques as microwaves, modulated light, rf transmission, magnetic-flux coupling, or by a carefully controlled direct connection. Magnetic-flux coupling is commonly used; however, it has several limitations. The drawbacks usually involve problems in cost and bandwidth. The direct connection is technically difficult but offers many advantages. The pros and cons of these techniques are discussed in Sections 5.11 and 5.16.

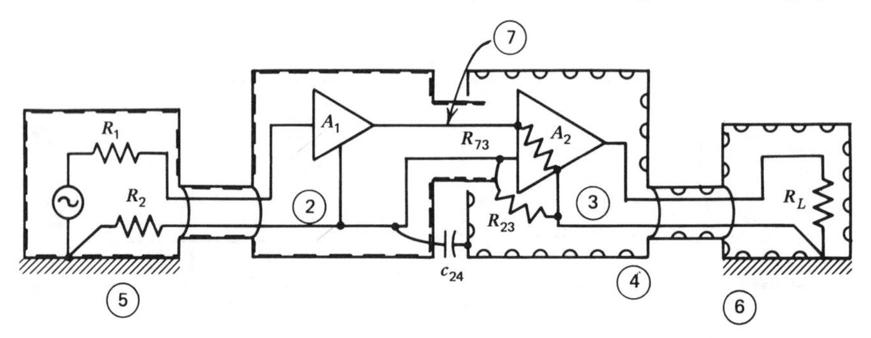

Figure 5.7c The direct connection method for differential amplifiers.

If magnetic-flux coupling is used, a modulation and demodulation scheme must be incorporated. A transformer is thus brought into play to transfer signal information in carrier form between shield enclosures. Figure 5.7*b* incorporates this elementary idea. Note that the modulator, transformer, and demodulator "straddle" the two shield enclosures. The dotted loop in Figure 5.7*b* represents the flux coupling between the signal regions. The modulation transformer contains the two signal shields. If the shield enclosure about the input signal is not complete, a leakage or mutual capacitance term such as c_{23}, together with any potential difference V_{14}, will permit current to flow in path ①→②→③→④→① and this current flows in the source resistance R_2. For a 120 dB CMR, this capacitance must provide a reactance of 1000 MΩ or greater.

The direct connection method involves conductors exiting the input shield enclosure in a controlled manner. If the common-mode rejection ratio requirement previously described is to be met, a 1000 MΩ path is all that is permitted. Since direct input connections to the amplifier within the second shield enclosure are necessary, this amplifier must have input impedances of 1000 MΩ or greater for both input connections. The direct-connection method is shown in Figure 5.7*c*.

A mutual capacitance such as c_{24}, together with a potential difference V_{56}, will permit current to flow in path ⑤→②→④→⑥→⑤. The input impedance R_{23} will permit current to flow in path ⑤→②→③→⑥→⑤. Examination reveals that these two paths are in parallel. c_{24} may be the result of cable or transformer shield leakage but R_{23} and R_{73} are built into the electronics of the instrument. They are therefore equivalent to built-in or controlled leakage paths. The design method for acquiring these high-input impedances involves feedback techniques and component knowledge and is not subject material for this book. The problems are difficult, but dc instruments have been built with input impedances at 60 Hz of 10,000 MΩ.

The input signal in Figures 5.7*b* and 5.7*c* is shown as a voltage source with two source resistors. These values must be considered for their total value and for their difference value (unbalance). If this unbalance is too high or if the total value is excessive the instrument's functioning limits may be exceeded.

The source as shown is connected to a ground point. This is typical of thermocouples, grounded strain gages, etc. The point in the transducer that is grounded is sometimes selectable, but often it is dictated by the process or problem. When a point along the signal path is connected to the external environment this point defines the zero-signal reference potential. By the rules previously developed, the input shield must make its connection to the zero-signal reference at this same point.

The high impedances required to reject common-mode signals are sometimes available at the input terminals of an instrument. It does not follow that this high input impedance permits high source-impedance operation. The specifications should be consulted if sources above a few thousand ohms are involved.

5.8 THE FLUX-COUPLED DIFFERENTIAL DC AMPLIFIER

General

The technique illustrated in Figure 5.7*b* does not indicate the division of gains in segments A_1 and A_2. These gains are a matter of design technique. Flux-coupled instruments have been built with all the gain in A_1 or all in A_2. Examples of both techniques are shown below.

5.9 INPUT MODULATOR TECHNIQUES WITH FLUX COUPLING

A 60 or 400 Hz modulator can be used to develop a carrier or ac signal proportional to the incoming signal. After a suitable input transformer, this carrier signal is amplified and then synchronously demodulated. The output information in carrier or demodulated form can be fed back for purposes of stabilization or input impedance enhancement. The carrier amplifier is A_2 and the modulator is the only element in A_1. Figure 5.9 shows a properly shielded instrument of this type without feedback.

The operation of the amplifier is as follows: The filtered input signal appears across capacitor C_1. The modulator M_1 impresses this potential on an alternating basis across the two halves of the input transformer. The ac signal on the transformer secondary is proportional to the signal on C_1, and the phase of this signal referred to the modulator drive is a measure of the input-signal polarity. The same signal that drives the modulator also demodulates the amplified carrier signal resulting in an amplified version of the input information.

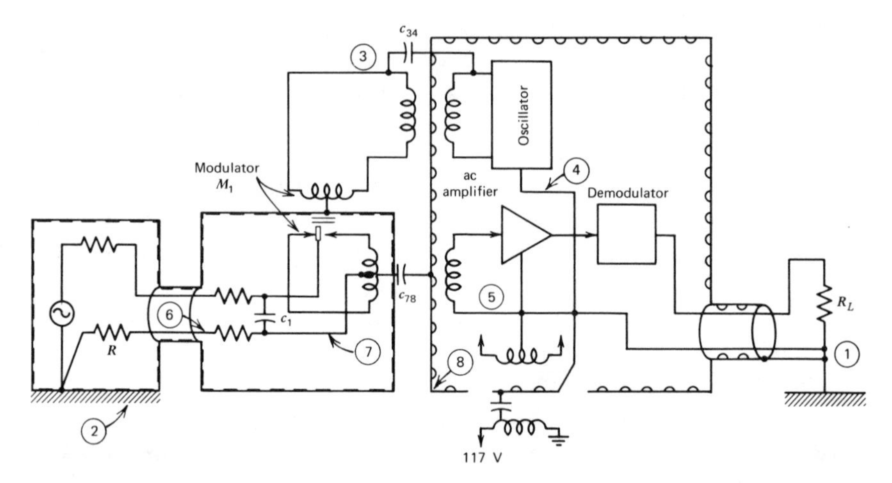

Figure 5.9 Input-modulator flux-coupled differential amplifier.

The shielding problems for this instrument involve three areas: the power transformer, the signal coupling transformer, and the modulator and demodulator drives. These three ac processes must each be considered separately. The problems that can be encountered are the same as those discussed in Chapter 4.

The power entrance for the carrier amplifier ideally requires a double-shielded power transformer. There are several reasons why this may not be provided. First, the power currents circulate principally in the output leads and are not subject to signal amplification. Second, the modulation frequency may be other than the power frequency. Under this condition there may be little or no effect caused by pickup.

The shielding of carrier potentials is important in this type of system. Any carrier signal entering the signal path incorrectly can cause offset or gain error. Since carrier current can circulate from transformer coils to shields and into signal conductors, these paths must be controlled. As an example, if the mutual capacitance c_{34} is excessive, then the oscillator drive V_{54} causes currents to circulate in the path ④→③→②→①→⑤→④ and this includes signal conductor ⑤.

Mutual capacitances such as c_{78} and any potential difference V_{12} permit common-mode currents to flow in the path ②→⑥→⑦→⑧→①→②. This current flows in the source resistance R and defines the CMR. By previous arguments, this capacitance should be kept below a few pF. If the primary coil of the input modulation transformer is not balanced, a signal carrier potential will also exist in series with c_{78} (this occurs because the mutual capacitance is effectively connected to a point removed from the midpoint). This carrier potential will circulate current in the same common-mode path as above. The harmful effects are probably minimum, but a good rule to follow keeps signal currents confined to signal conductors. A third shield placed next to the primary coil and connected to ⑦ will eliminate this effect.

5.10 POSTMODULATOR TECHNIQUES IN FLUX-COUPLED INSTRUMENTS

The instrument in the preceding section has specifications limited by the input modulator. When the flux coupling is placed nearer the output in gain level, modulation errors are not amplified by the full gain. This permits a modulation technique that optimizes a new set of parameters. For example, a postmodulator permits high-frequency solid-state switching, and this can provide greater bandwidth. Drift in such a modulator is not further amplified and can therefore be proportionately greater. An example of this type of instrument is shown in Figure 5.10. Note that the gain following the demodulator is unity. This technique can provide an upper frequency response in excess of 20 kHz. The power for the input amplifier A_1 can be from a separate power transformer or as shown from the oscillator on the A_2 side.

The power transformer is shown double-shielded although a single shield would be adequate in most installations. The transformer's coupling

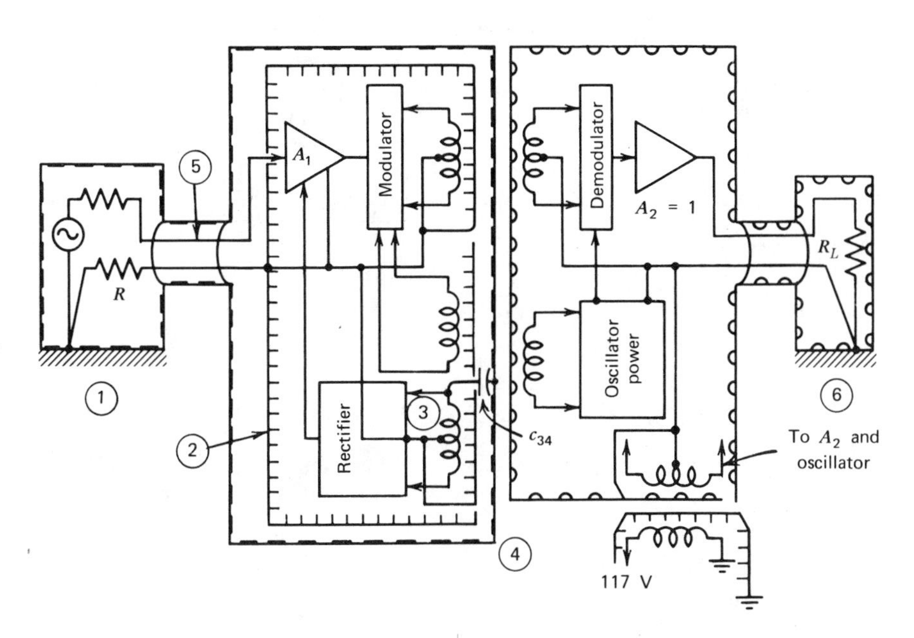

Figure 5.10 A postmodulator flux-coupled differential dc amplifier.

of carrier for power or signal must be carefully considered. If a mutual capacitance term such as c_{34} is excessive, a potential V_{23} can cause carrier current to flow in the path ①→②→③→④→① and this path includes the line-unbalance element R as well as the input-signal conductor. Carrier that reenters the A_1 amplifier can cause gain errors or regeneration problems. Input filters can be used but this reduces the input impedance specification. The best compromise involves bandwidth limiting in the amplifier and shielding to reduce mutual capacitance effects such as c_{34}. Shielding carrier transformers is difficult because of the high frequencies and square-wave drive signals that are often used. Conventional shields can be avoided by techniques as discussed in Section 8.17.

The input conductors ⑤ and ② in Figure 5.10 are nonsymmetrical; a symmetrical form is shown in Figure 5.12. An amplifier that has both input conductors treated equally is referred to as a *true*-differential input amplifier. Symmetry requires that both input leads have the same input impedance relative to both zero reference conductors. The word impedance is stressed here. It is not sufficient to have the dc resistances equal. Since some applications require an unsymmetrical instrument, the problems that are presented are worth discussion.

In Figure 5.10 input conductor ② is also one of the shields in the modulation transformers. This places a shield to zero-signal capacitance between signal conductor ② and conductor ⑥. Note that signal conductor ⑤ does not share such a capacitance.

The presence of an unbalance capacitance as described above permits an unsymmetrical flow of common-mode current in the input lines. Depending on cable length, frequency of interest, etc., this unbalance can reduce performance.

Balancing the input by adding fixed capacitors to the input line is sometimes tried. This can only result in reducing the common-mode rejection ratio. Further, since active capacitances are a function of feedback, any reactive balancing will be gain- and temperature-dependent. Since the feedback factors vary with the gain elements used, reactive balancing must be a hand-tailored process.

5.11 MERITS OF FLUX-COUPLED INSTRUMENTS

The flux-coupled differential dc amplifiers have several advantages. These are:

1. Common-mode potentials are placed across transformer shields and thus the maximum common-mode level is minimized by transformer insulation resistance.

2. Floating sources are permitted; that is, the circuitry within the input shield functions without regard to external potentials.

3. Common-mode signals are not actively processed and therefore there are no frequency- or amplitude-related factors (slewing-rate limitations). Common-mode rejection factors are independent of gain when the post-amplifier gain is unity.

4. The input modulator system requires no electronics in the input shield enclosure. Input signal and shield currents are not influenced by a power entrance into the input shield enclosure.

The disadvantages of the flux-coupled differential dc amplifiers are:

1. Over-all feedback processes require a second flux coupling. This means performance is limited to the quality of a flux-coupling mechanism.
2. Bandwidth is limited by the modulation frequency.
3. Carrier currents can exit the instrument via both the input and output shields if the transformers are not perfectly balanced.
4. The input shield must be reactively associated with the output shield at the flux-coupling interface. This shield-to-shield capacitance permits shield currents to flow. If these shield currents should flow in part of the source impedance, the effect can be common-mode to normal-mode conversion. See comments on medical instrumentation, Section 6.13.
5. Unsymmetrical input instruments have frequency response limitations dependent on line unbalance.
6. Source resistances may be restricted to value of 1000 Ω or less unless input filtering is used.

5.12 THE ELECTRONICALLY COUPLED DIFFERENTIAL DC AMPLIFIER

Electronic coupling can be used with any gain division between A_1 and A_2 (see Figure 5.7*c*). It is common practice to place all of the gain in A_2 but occasionally this is not an optimum choice. The problem of rejecting the common-mode signal is directly related to the electronic coupling required. As pointed out in Section 5.3, the coupling impedances should be in excess of 1000 MΩ.

A powerful technique for accommodating high-voltage common-mode signals is available if the output from amplifier A_1 is coupled to amplifier A_2 through a balanced attenuator. This technique attenuates both common-mode signals and normal-mode signals in the coupling attenuator. Amplifier A_1 can then be used to reject the common-mode content and reamplify the signal. If the common-mode level is 1000 V, then a 33:1 attenuator reduces this level to 30 V within the capabilities of amplifier A_2. This technique is outlined in Figure 5.12. Power for amplifier A_1 must be triply shielded.

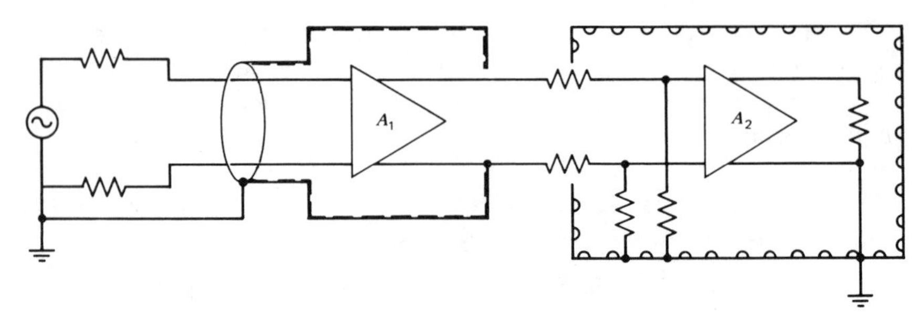

Figure 5.12 A high-voltage common-mode scheme.

Note that ground potential differences force current to flow in the attenuator. The impedance in this path is usually about 150 kΩ. The resulting input shield current can couple signal into A_1 at ac if the input cable is very long. There is no coupling at dc if the input cable is free from leakage. The attenuator resistors must be selected very carefully for dynamic balance because the quality of this arrangement is often limited by the quality of the attenuation.

Current flowing in the coupling attenuator does not affect the common-mode rejection ratio. This specification is limited by the input impedance of amplifier A_1. The impedance provided by the attenuator is seen by the common-mode source. As long as this current does not flow in a source resistance, no signal contamination will result.

5.13 POSTAMPLIFICATION IN ELECTRONICALLY COUPLED INSTRUMENTS

Figure 5.13 shows an instrument where the gain resides entirely in A_2. This specific circuitry is not shown for its practicality but for function only. Circuits with more sophisticated feedback techniques are usually required to meet specification demands. The input transducer is shown grounded. By Rules 1 and 2, the input shield is connected to the zero-signal reference potential at this point. It is of interest to note that the instrument requires this shield potential but that the shield makes *no* connection within the instrument. *This is correct;* the shield should be connected to zero-signal reference potential once and once only and this is properly done at the source as shown. In this example no electronics appears on the A_1 side.

The mutual capacitance c_{24} represents leakage out of the input shield. This capacitance is in parallel with the input impedance Z of the instrument. These reactive paths need not be the same for the two input lines as the input unbalance $R_2 - R_1$ senses the magnitude of the total current flow. This results because the shunt capacitance of even a short

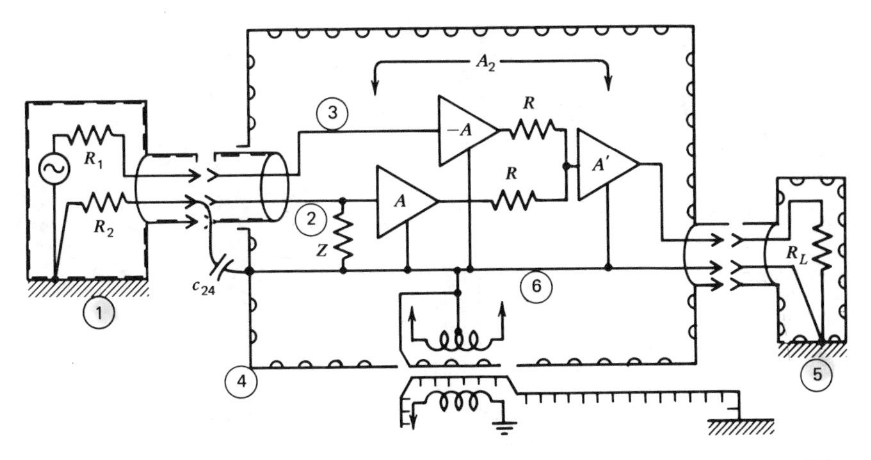

Figure 5.13 An electronically coupled differential dc amplifier with postamplification.

cable connection parallels the two current paths.[1] The paralleled input impedance should be in excess of 1000 MΩ to meet the conditions outlined in Section 5.3.

5.14 PREAMPLIFICATION IN ELECTRONICALLY COUPLED INSTRUMENTS

The preamplification technique in electronically coupled differential instruments is similar to that used in the flux-coupled instrument shown in Figure 5.10. Power to the A_1 side can be from a special modulator, from a separate power transformer, or from a common power transformer as shown in Figure 5.14.

The preamplifier is single ended with a local shield ③ segmented within the power transformer. If this shield were omitted, any potential difference V_{24} and mutual capacitance c_{45} would permit current to flow in the loop ②→④→⑤→①→②, which includes the source resistor R. This local shield is required by Rule 1 and is connected to zero-signal reference potential at the amplifier input. It should be noted that the zero-signal reference for the amplifier does not correspond to the zero-signal reference for the signal. The resistor R connects these two reference points. It is obvious that only signal currents are permitted to flow in this resistance. To insure proper shielding, three shield regions must be penetrated for power to enter the input enclosure. Two such regions must be penetrated for the output enclosure.

Common-mode rejection capability is dependent on all paths that

[1] This shunting effect is frequency dependent and obviously has no effect at dc.

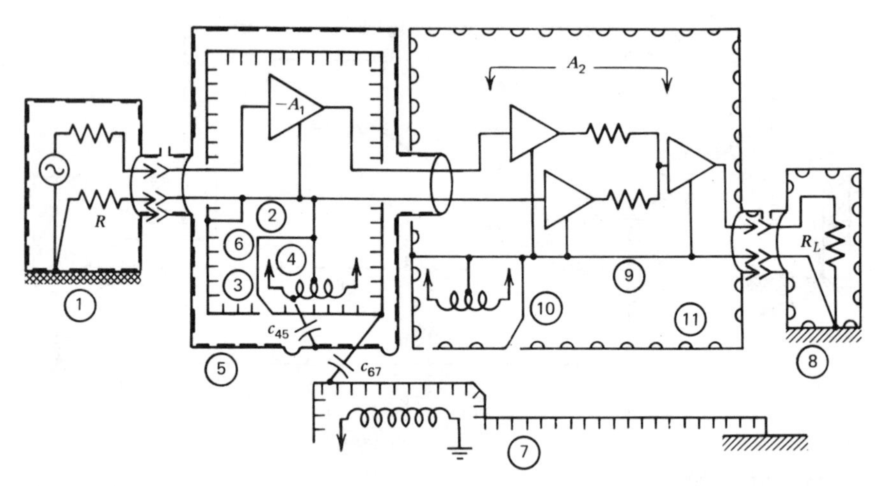

Figure 5.14 An electronically coupled differential dc amplifier with preamplification.

permit current to leave the input shield enclosure. The input impedances to the A_2 section must be high for the same reasons given for the post-amplifier discussion. The input shield must be well controlled for these same reasons. A new source of difficulty has been added however. This occurs when the input shield is in proximity to the primary shield of the power transformer. Any openings here will allow common-mode current to flow to power earth. Referring to Figure 5.14, potential difference V_{17} and the capacitance c_{67} permits current to flow in path ⑥→⑦→①→②→⑥ and this includes the source resistance R.

The primary shield ⑦ should be at or near the potential of the ground point ⑧. If not, any potential difference V_{78} can cause unwanted current to flow in conductor ⑨ via the path ⑦→⑩→⑨→⑧→⑦. If the primary shield is at the same potential as the input ground point ①, any potential difference V_{75} causes current to flow in the input shield path ⑦→⑤→①→⑦. This may be undesirable where long input cables are required. If this connection is used, however, the current in conductor ⑨ could be eliminated by adding a third shield between conductors ⑦ and ⑩ that is tied to conductor ⑪. This further complicates the transformer design but it does insure that all currents are contained in shields, not in signal conductors.

5.15 RETURN-PATH REQUIREMENTS IN ELECTRONICALLY COUPLED INSTRUMENTS

All electronically coupled differential amplifiers rely on high input impedance amplifier designs for their performance. This high impedance defines the common-mode rejection capability for the instrument. High input impedance and operation from high source impedances are not necessarily related. In many designs, the source impedance may be restricted to 1000 or 2000 Ω. The total source resistance for the amplifiers in A_2 is the signal resistance R_1 or R_2 plus the resistance between the input and output ground connections. In Figure 5.13 this is resistance R_2 plus the resistance between ground points ① and ⑤. This sum is called the return-path resistance for the amplifier input. If the input circuit is floating, that is, if no ground tie is made, the return path for conductor ② is just the leakage resistance of cables and connectors and this value is usually unknown and unpredictable.

A high-value return path for the amplifiers in A_2 places a design limit to the permitted operating currents (base or gate currents) in these connecting conductors. Even if these currents can be kept below 10^{-10} A, a leakage resistance of 10^{12} Ω would be unacceptable. This current level in 10^{12} Ω will cause the amplifiers to be blocked by their own common-mode operating voltage. This implies that there must be a return path considerably less than 10^{12} Ω in the system. More about floating source operation is given in Section 6.9.

5.16 MERITS OF ELECTRONICALLY COUPLED INSTRUMENTS

Electronically coupled instruments can be designed to meet most instrumentation requirements. The specific advantages are listed below:

1. The coupling processes can be quite accurate and stable without the use of over-all feedback.
2. Bandwidth is not limited by a modulation process.
3. The postamplification technique requires no power entrance into the input-signal enclosure. This eliminates the possibility of power-current flow in the input shield.
4. The postamplifier techniques permit a completely symmetrical input (true differential).
5. The capacitance between the input shield and all other shields or ground conductors can be kept to a minimum (postamplifier techniques).

The disadvantages are listed below:

1. The common-mode signal must be processed by the circuitry. Since these signals are processed together with normal signals, this imposes a difficult design requirement.
2. Common-mode signals are restricted in magnitude by design. Special techniques are required to handle more than 20 V.
3. Common-mode rejection is frequency dependent as well as frequency-level dependent.

4. A return path for the input signal and its enclosure must be guaranteed somewhere in the system. (This can be provided within the instrument. See Section 6.9.)

5.17 PHOTOCOUPLING TECHNIQUES

A photocoupled transistor can be used to couple signals between amplifiers A_1 and A_2. The nonlinearity in this transmission can be balanced out by using a matching photocoupler element in the feedback loop of the driving amplifier. The performance of this arrangement is limited by the matching that can be obtained between photocouplers. Common-mode signal level is limited by nature of the ic substrate. Best matching requires that both photocouplers be on the same substrate. Common-mode rejection ratios are limited by the amount of guarding that can be placed between the couplers.

6

General Application Problems

A FEW EXPLANATIONS

The figures used in this text show shield lines completely enclosing all signal conductors. In practice this does not occur. For example, an input element may be "out in the open." However, because it is connected and associated with a zero-signal potential, it is in effect protected by that potential. The shield enclosure in this environment is only symbolic and need not be taken completely literally. In regions of high impedance, or where small signals are involved, the complete shield may actually be required. It should also be noted that elements, instruments, and cables may be physically arranged in many ways. The process described are typical and the diagrams are only representative. The concepts are important and they can be applied regardless of the specific component organization.

Symbols for ground and earths are varied as they are in the literature. Again it must be pointed out that a ground may not be ohmically connected to earth (dc). Power neutral is almost always earthed and is indicated by the symbol ⏚.

6.1 WHEN SINGLE-ENDED AMPLIFIERS SHOULD BE USED

A single-ended instrument was defined to be a device with a single zero-signal reference conductor for both the input and output signals. By the rules given previously, this conductor can be connected at a single point to an external ground. If these conditions prevail, a single-ended amplifier can be applied.

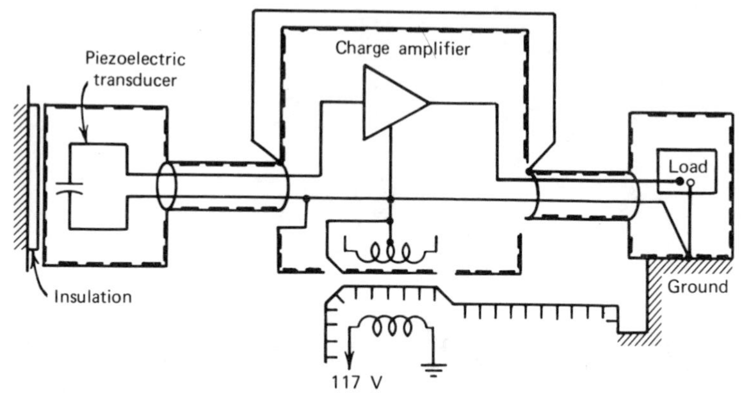

Figure 6.2 A single-ended charge amplifier and a grounded output.

6.2 CHARGE AMPLIFIERS

Charge and voltage instrumentation used with piezoelectric transducers are usually single-ended in nature. Here the crystal transducer at the input is shielded by a metal case that is insulated from the crystal element. An insulating block is usually placed between the transducer and the structure being tested to avoid any improper shield connections. This arrangement is shown in Figure 6.2.

The transducer insulation shown in Figure 6.2 poses a problem if the mass that is added changes the dynamics of the system being measured. The transducer can be made lighter if external and internal insulation is not required. Both these insulation needs are unnecessary if a "two-ground" instrument is used. With this type of instrument, the transducer can be "grounded."

The conversion of charge to voltage in any charge amplifier is a single-ended process. The signal generation in most charge transducers is also single-ended (unbalanced). This means that balanced techniques such as in Figure 5.13 are not allowed. A permitted technique parallels the circuitry shown in Figure 5.14 where A_1 is a charge converter, the bottom resistance R is zero, and the top source resistance is replaced by a capacitance equal to the transducer value. A charge amplifier that can be operated between two grounds is called a differential charge amplifier.[1]

6.3 INPUT COAX APPLIED TO A SINGLE-ENDED AMPLIFIER

The charge or voltage amplifier in Figure 6.2 has applications where the transducer may be in a high acoustic field or in a region of intense vibration. The input cable in the vicinity of the transducer can be a

[1] It is always possible to cascade a single-ended charge amplifier and a low-gain differential amplifier. Except for the added expense this is a permitted solution.

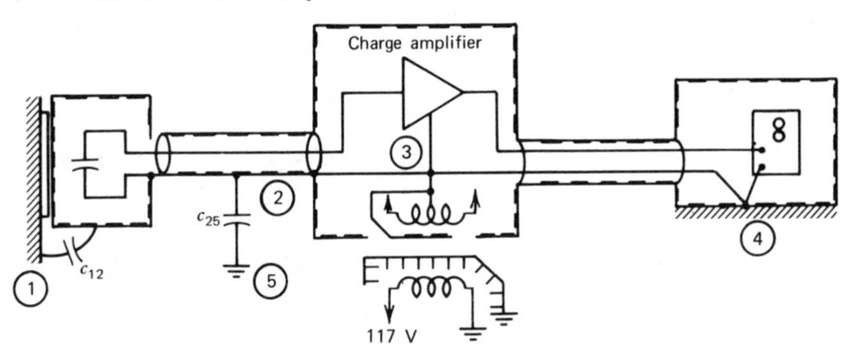

Figure 6.3 Use of coax in a single-ended amplifier.

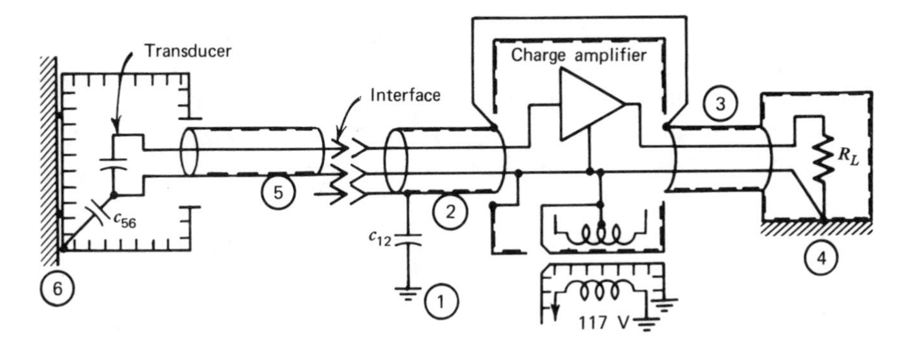

Figure 6.4 Coax to twinax interface on an input line.

serious noise generator if special cable is not used. (See Section 9.20.) If coax is used the shielding configuration of Figure 6.2 is violated. Since the outer conductor of the coax is a signal conductor, currents that are parasitically picked up by this conductor are carried the entire length of the signal conductor to the earth point. This configuration is shown in Figure 6.3.

Any potential difference V_{14} or V_{45} causes currents to flow in loops ①→②→③→④→① and ⑤→②→③→④→⑤ through the capacitances c_{12} and c_{25}. These currents cause noise pickup in conductors ② and ③ depending on the length of input cable involved.

6.4 COAX-TO-TWINAX INTERFACE

If coax must be used at the transducer, a crossover to twinax can be made at a convenient point. This eliminates some of the parasitic pickup phenomena. If the case can be floated from the coax this is also a help in reducing unwanted pickup. Figure 6.4 shows a practical configuration.

A potential difference V_{14} will cause current to flow in capacitance c_{12} in the loop ①→②→③→④→①, a path involving shields only. If the coax line ⑤ can be kept short and if the transducer element has a small capacitance c_{56} to its case, this configuration can provide good immunity from noise contamination caused by unwanted current flow in signal conductors.

6.5 RESISTANCE-BRIDGE APPLICATIONS (STRAIN GAGES)

Single-ended instrumentation is often used to amplify resistance-bridge signals. Bridge excitation can be by battery or by special floating power supplies as discussed in Chapter 7. If the bridge is ohmically connected to a structure, then no other connection of the zero-signal reference conductor is permitted. This implies that the load must be floating, i.e., the load must be a resistor or meter or galvanometer. The grounded bridge permits the structure being measured to be a part of the shield closure around the signal at the bridge. This configuration (with a battery used for bridge excitation) is shown in Figure 6.5.

6.6 ISOLATED RESISTANCE BRIDGE AND A GROUNDED OBSERVATION POINT

If the bridge in Figure 6.5 is ohmically free from a ground connection, then a ground connection is permitted at the instrumentation output. This configuration is not free of difficulty as the circuit in Figure 6.6 indicates. Here structure ① represents an opening in an otherwise tight signal shield. This structure ① is not necessarily at the zero-signal reference potential established at ④. The potential difference V_{14} will cause cur-

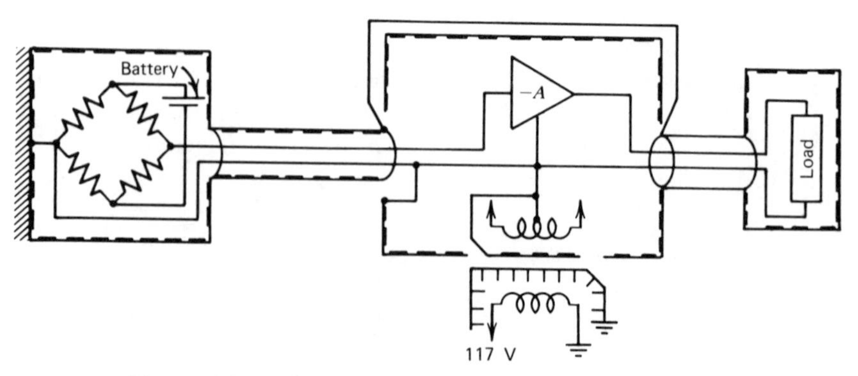

Figure 6.5 A single-ended amplifier and a grounded bridge.

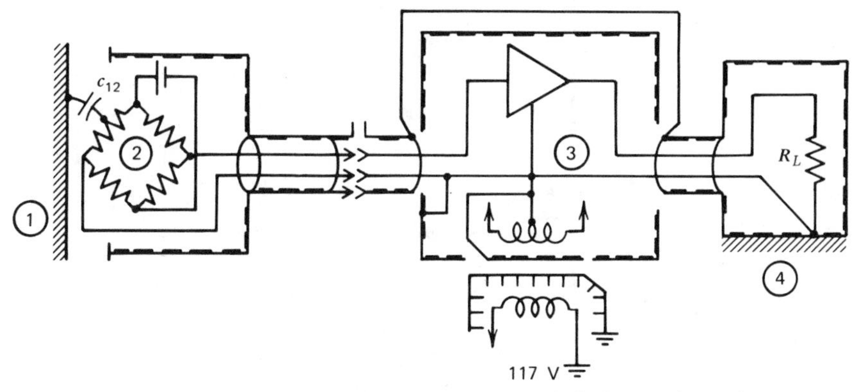

Figure 6.6 A floating bridge and a grounded observation point.

rents to flow in parasitic capacitances such as c_{12} in the loop ①→②→③→④→①. Since this current can flow in bridge resistances, large noise signals can be picked up.

If the capacitances to ① from all bridge arms are equal and the bridge is balanced, then at first glance the noise potential difference V_{14} is balanced out by the bridge. The capacitances involved are usually hundreds of picofarads, and it is unlikely that these effects will be completely balanced unless special precautions are taken. Many applications arise when part of the bridge is made up of elements that are not located at the structure under test. In this situation the reactive unbalance can be much greater and the configuration of Figure 6.6 can be impractical.

The strain-gage power supply in Figure 6.6 is a battery. In practice floating power supplies are used. A full discussion of these power sources and application to resistance bridge instrumentation is given in Chapter 7.

6.7 SINGLE-ENDED AMPLIFIERS AND THERMOCOUPLES

Thermocouples and single-ended amplifiers are not usually recommended in combination because of the noise pickup that usually results. This noise is attributable to the input line resistances required in thermocouple instrumentation. This noise can be reduced by using unbonded elements or by restricting the bandwidth of the instrumentation to 10 Hz or less.

Bonded or grounded thermocouples are the most difficult problem to treat. The resistance R_2 between the zero-signal reference conductor and the input ground presents the difficulty. This condition is shown in

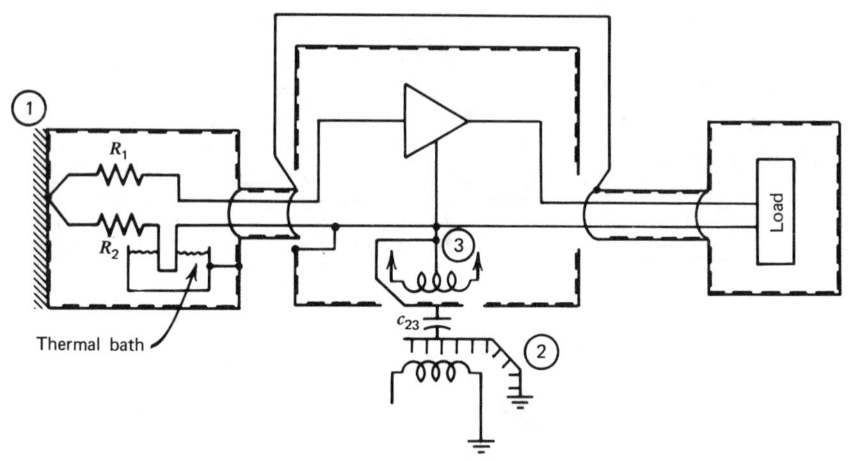

Figure 6.7 A bonded thermocouple and a single-ended amplifier.

Figure 6.7. A potential difference between grounds ① and ② causes currents to flow in the loop ①→③→②→① through capacitance c_{23}. This current flows in line resistance R_2. This resistance can be several hundred ohms, much higher than that present in an input cable run. This condition simply requires that currents in R_2 must be kept correspondingly low or, if these currents do flow, that the amplifier ignore these signals by being bandwidth limited.

If an unbonded thermocouple is used and the output signal line is grounded, the situation is exactly the same as shown in Figure 6.6 except that the element is self-generating and needs no battery. The pickup problems caused by thermocouple capacitance to the structure are identical. Here, however, the source is unbalanced and input symmetry is not available for first-order cancellation of unwanted signal pickup as in the case of the bridge circuit.

Thermocouple signals can be amplified by using differential amplifiers. The treatment of bonded and unbonded elements varies somewhat and is treated in Sections 6.9 to 6.11.

6.8 WHEN DIFFERENTIAL AMPLIFIERS SHOULD BE USED

Chapter 5 dealt with the essential nature of the differential amplifier. An instrument that can amplify signals developed in one zero-signal reference environment and present them in conditioned form to a second zero-signal reference environment was defined as a differential amplifier. Applications where this need exists are numerous. These include thermocouples, strain gages, and data-transfer links.

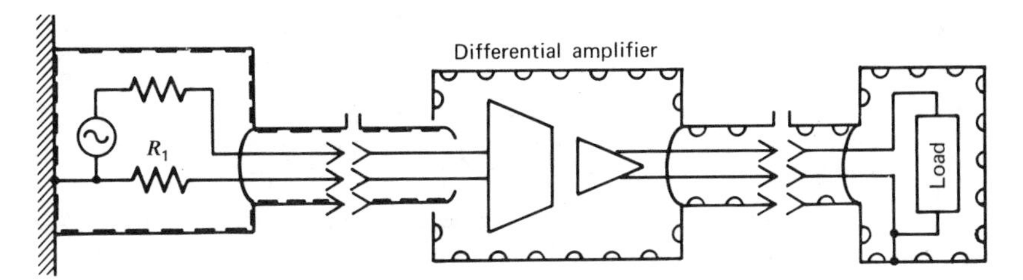

Figure 6.8 A single ground point where a differential amplifier is required.

The need for differential amplifiers is often stated in terms of the problem that must be solved. For example, if the source and output terminating points must be kept ohmically disconnected, or if resistive ties to ground at the source are required, differential amplifiers are indicated.

Figure 5.13 shows an application where two ground reference points are dictated. Here the isolation between signal regions may be required to be as high as 1000 MΩ. The source might be a thermocouple and the output might be a tape recorder or computer.

Another application that usually requires a differential amplifier approach is shown in Figure 6.8. Here a resistance R_1 lies between the input signal conductor and the input ground connection. Single-ended instruments can be used but designs usually pay little attention to operation under this condition.

For values of R_1 above 20 or 30 Ω, a differential amplifier is recommended. There is an assumption here that bandwidth is required and that filtering is not a permitted solution to avoiding the pickup problem.

6.9 FLOATING SOURCES AND DIFFERENTIAL AMPLIFIERS

The floating-input problem was discussed in part in Section 5.15. The ground-return problem was not a serious problem for flux-coupled instruments but had to be considered if the coupling was electronic. This problem arose because these instruments required an input current return path through the grounds. Without this path, the input circuits would block up because of self-generated dc common-mode signals.

The term ground needs to be clarified here. Ground, as defined earlier, refers to the zero-signal reference of a complex of electronics. The complex is usually operated from utility power via a power transformer. This complex may or may not be ohmically connected to earth. Grounds and earth connections are not synonymous. An oscilloscope is a ground point but it is not necessarily earthed unless the third pin on the power

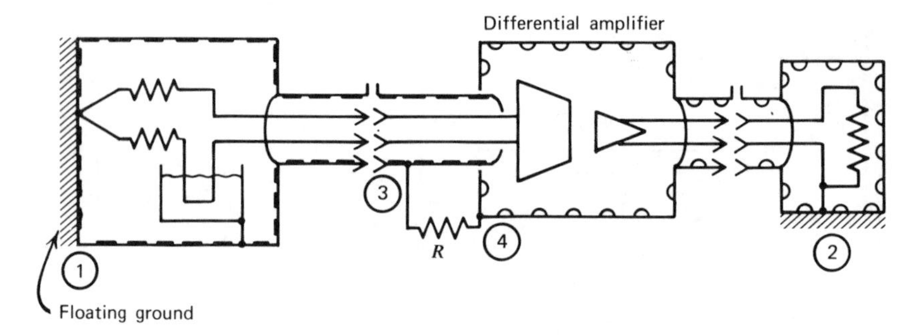

Figure 6.9 A return path solution in differential amplifier application.

cord is used. These statements imply that grounding both zero-signal reference conductors in a differential amplifier is not sufficient to guarantee that a ground return path exists. The grounds must be ohmically connected at dc to fulfill the return-path requirement.

If the ground return path for the input and output circuits of a differential amplifier is not provided, a return-path guarantee can be placed into the system. The solution is shown in Figure 6.9.

In this example, the input ① is grounded but *not earthed*. The ground return path occurs only through resistor R. A potential difference between ① and ② (common-mode signal) causes currents to flow in the path ①→③→④→②→①. This current flows only in shields and grounds and does not enter a signal conductor. This solution is valid as long as R can be kept moderately high. If this resistor permits large shield currents to flow, the resulting potential drop along the shield can couple into the signal processes. (A shield, to be effective, should be at zero-signal reference potential along its entire length. An unwanted shield difference of potential violates this objective.)

The resistor R in Figure 6.9 is sometimes a part of an instrument's design. If it is not present it can be easily placed into the system externally. Its maximum value can be determined from the instrument's specifications on "pump-out" or input current if no other specification is apparent.

If this resistor is to be high, say 10 MΩ, then the instrument must be capable of operating from moderately high source resistances. It is wise to caution against a common misconception. High input-impedance and high source-impedance operations are not synonymous. The use of high values for R in Figures 6.9 is not necessarily available in all instrument designs.

Resistor R does *not* affect common-mode rejection. This specification is controlled by the input impedances to the electronics. Resistor R also

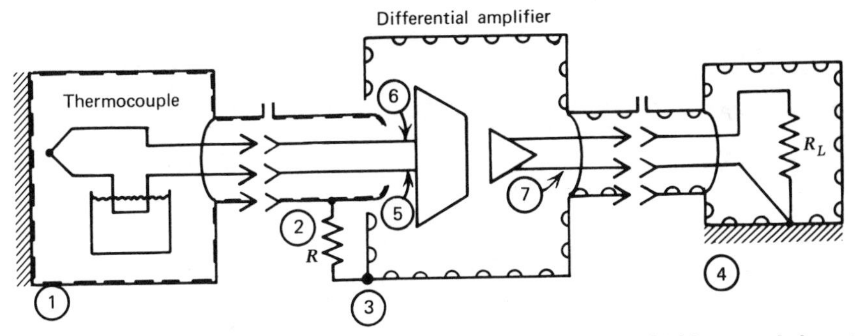

Figure 6.10 An unbonded thermocouple with the input shield grounded—*not* recommended.

does *not* affect the input impedance to the instrument. Resistor R does permit some shield current to flow, but under normal conditions shield currents of this same magnitude are easily picked up parasitically.

Designers often shunt R with a series resistor and capacitor. At high frequencies the value of R is reduced to a new value R'. This arrangement is used to reduce the common-mode content to levels that can be handled by the instrumentation. If R' is about 100 Ω it acts as a termination to the earth-shield transmission line. See Section 9.12 for further discussion.

6.10 A *MISAPPLIED* SHIELD FOR THERMOCOUPLES

The question is often asked, "Can the input shield be grounded but the signal conductor left disconnected from the shield?" This situation is shown in Figure 6.10. One obvious defect here is that there is no ground return path for signal lines ⑤ and ⑥. This is not acceptable if the amplifier is electronically coupled. The resistor R between points ② and ③ does not provide this path. If a path exists, it is the leakage resistance within the input cable.

A resistor placed between ⑤ or ⑥ and ⑦ does provide a return path but this violates the input impedance and therefore spoils the common-mode rejection capability.

Flux-coupled instruments will often tolerate the situation shown in Figure 6.10 but it is still not recommended. The first rule of shielding is violated with this arrangement. The potential difference between the signal conductors and the shield is undefined and can lead to instrument malfunction.

6.11 THE DOUBLE INPUT SHIELD FOR THERMOCOUPLES

If the input shield is not grounded at the source and if the thermocouple is not bonded, an extra shield can be used to reduce noise pickup. The

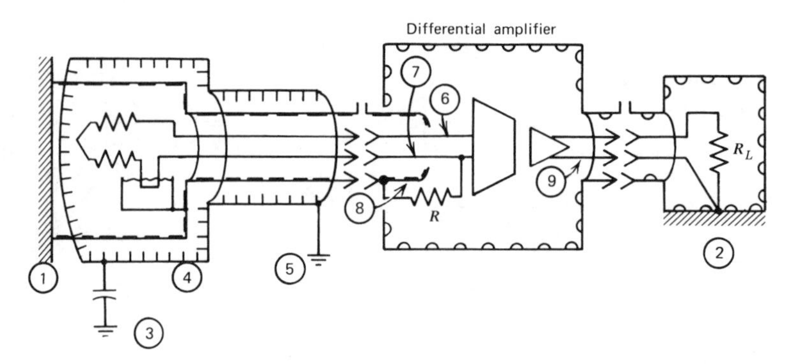

Figure 6.11 A double input shield for ungrounded thermocouples.

solution is shown in Figure 6.11. The structure ① is connected to the inner shield and thus provides shield closure at the thermocouple. The outer shield ④ is connected to ground at ⑤. Grounding this outer shield at the instrument may be one of the parameters dictated for safe or proper operation. The floating thermocouple by itself does not provide a ground return path for conductors ⑥ and ⑦. This is provided in resistor R placed between conductor ⑥ or ⑦ to ⑧ . The input return path is now ⑦→⑧→①→②→⑨. Resistor R can be as high as 10 MΩ depending on the input-current levels for the instrument.

If the thermocouple should contact the structure at any time the resistor R would shunt some fraction of the thermocouple. The effect on performance would be negligible if R is high enough in value. If the thermocouple accidentally bonds, the shield configuration is identical to that shown in Figure 5.9 except that an extra outer shield is present. Obviously, if the thermocouple bonds at two points the element is defective. If the return path is not provided, the element may be operative but the instrumentation may not be.

The floating thermocouple should be viewed with caution. Care should be taken so that the instrumentation is compatible with this mode of operation. The first rule of shielding requires that the shield be connected to the zero-signal reference point. Just grounding the shield does *not* satisfy this requirement.

6.12 GROUNDING VS FLOATING SIGNAL LINES

The distinction between grounding and earthing has been pointed out in Section 6.9. An entire system of ground connections can be floated and not involved with earth at all, that is, in an aircraft, spacecraft, or even in a laboratory. The potential difference between this system of grounds and earth can be of concern from a safety standpoint. It is accepted practice to make such a connection once to a conductor buried in the earth. See Chapter 10 for further discussion.

The multiple shield potentials described in previous sections arise because grounds and earths support current flow. Shorting points together is usually a fruitless effort. The proper solution uses the instrumentation to ignore or process out these differences. The use of floating sources discussed in the previous sections raises the issue of whether these points should be earthed or permitted to float. A firm answer cannot always be given but some suggestions can be made.

Differential instruments are designed to operate with two zero-signal reference potentials. To operate the instrumentation with only one point grounded leaves the other point free to assume some average potential determined by parasitic processes. It is safer and more reliable to define this point rather than leave it to find its own value.

If the floating input line to a differential amplifier should contact an unknown conductor, the performance might still be adequate but the system becomes an unknown entity. If this contact is high voltage, someone might be in for a rude shock upon touching a supposed signal shield. If the signal is grounded, a second contact would cause a malfunction which should then be easily detectable. If a choice exists, an earthed signal line is recommended. If circumstances should dictate a floating signal, safety and instrumentation performance should be considered.

One common practice that provides a return path for pump-out current is to add a so-called Wagner ground, as shown in Figure 6.12. This circuit adds a balanced connection between the input circuitry and the output ground. A high-value resistor R_1 limits the common-mode current. This circuitry forces the average input common-mode signal to zero. Obviously if the active arms of the bridge are not reactively balanced

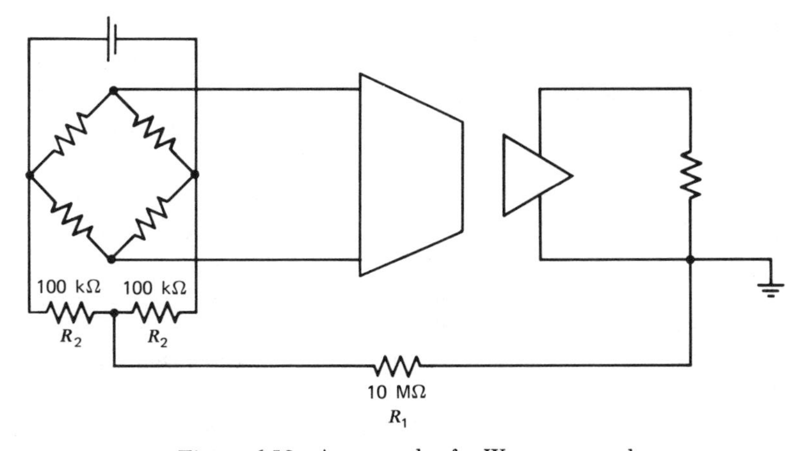

Figure 6.12 An example of a Wagner ground.

with respect to a contaminating source ground, high-frequency common-mode current can flow in R_1. Some designers raise R_2 to 1 MΩ and set R_1 equal to zero. This is basically the same Wagner ground.

Thermocouples are not generally balanced sources. In these applications designers often place the two resistors R_2 across the input leads of the amplifier and use R_1 to provide a pump-out current return path. Again the dc problems are resolved but only at the expense of poor high-frequency common-mode rejection ratios.

6.13 SHIELD-CURRENT CONTROL (THE MEDICAL PROBLEM)

The signals available on the body for instrumentation have relatively high source impedances. There is no one point on the body which can properly be called the best zero-signal reference. Any point will suffice although there may be a valid reason for selecting one point over another. Once this point is selected, then resistances exist to all other points used for signal sensing.

The currents that flow in the body are either self-generated or they result from parasitic involvement with the environment. If conductors are attached to the body a new distribution of current flow takes place. If the new currents that flow stem from external sources there is an excellent chance they will be sensed as normal signals.

The simplest process involves connecting one ground conductor to the body. The difficulty this creates is shown in Figure 6.13*a*. Potential differences such as V_{45} and V_{14} cause currents to flow in capacitances c_{57} and c_{13} through resistances R_1 and R_2. These resistances and capacitances are distributed in three dimensions, but the effect of observing signals, say between points ③ and ④, is the same as if discrete components were involved.

If the ground point ④ were removed the currents could still flow but they would probably be much smaller. If a ground connection is made, it is obvious that potential differences such as V_{14}, V_{54}, and V_{84} should be kept as small as is possible to eliminate this current flow. If the connection ④ is made via some electronic apparatus, the effects described in Chapter 4, Figure 4.12 become important. Here transformer-coil potentials can cause 60-Hz currents to flow. Under this condition no signal points on the body will be free from this 60-Hz contamination.

The requirement to ground the zero-signal reference in Section 6.12 seems to cause difficulty here. This condition is shown in Figure 6.13*b* for a typical signal-sensing arrangement. Currents caused by potential difference V_{12} take the path ①→④→⑤→②→① or ①→③→⑤→②→①. These currents cause a potential drop in R_{45} and this is a normal signal

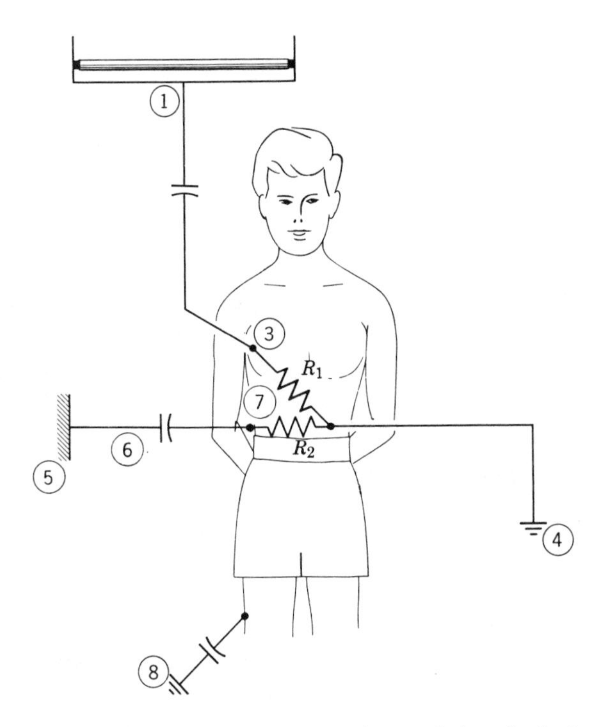

Figure 6.13*a* A single ground conductor tied to the body.

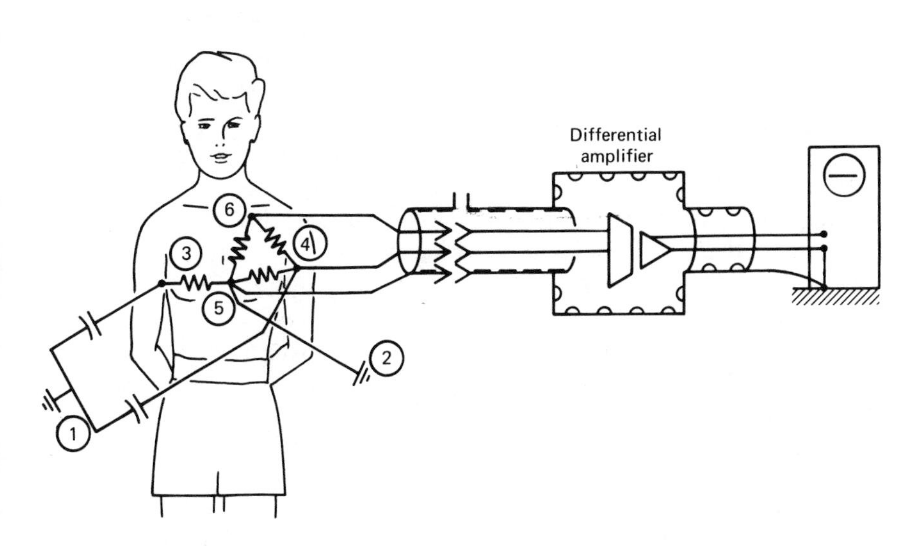

Figure 6.13*b* Typical signal sensing on a grounded body.

for the amplifier. If ground ② is removed, the unwanted pickup is usually less by virtue of the reduced current flow.

When the signal is sensed between specific points on the body, hum pickup can be very small. In the Figure 6.13*b*, this might be between points ③ and ④. Here the ground point is symmetrically located with respect to the external pickup so that $V_{35} = V_{45}$. Arrangements that require a special ground point to cancel spurious effects are not desirable.

If the instrumentation in Figure 6.13*b* operates without an input ground connection to ②, a ground return path for the instrument may have to be provided elsewhere. This path can be added between shields if not available within the instrument. See Figure 6.9.

6.14 THE USE OF ISOLATION TRANSFORMERS

Multiple-shielded power transformers (isolation transformers) are often used in instrumentation. If this application is to be meaningful the processes of isolation should be carefully considered or the transformer may be ineffective.

Much of the instrumentation that is commercially available has single shielding in its power transformers. The hope would be to add a second or primary shield by adding one isolation transformer to the system. Presumably the lack of shielding in the instrumentation would then be overcome. Although this is a commendable motive, the isolating transformer may not solve the system problem. It may only change the problem as indicated in Figure 6.14*a*. If the secondary coil of the isolation transformer is not grounded, potential differences between grounds ① and ⑥ can circulate currents in the path ①→②→③→④→⑤→⑥→①, and this path enters the zero-signal reference conductors for both instruments. (The process is complicated by other capacitances but this loop is still present.) The secondary coil of the isolating transformer must be grounded to avoid this loop. It is important to note that these processes are not dependent on the shield connection ⑦.

The choices for grounding the secondary coil are obvious. Either side of the coil, or perhaps the center tap ⑧, can be grounded. If either side is grounded the basic problem of single-shielded transformers is presented. See Section 4.12. If the center tap is grounded to ① at least the potential causing the current flow in amplifier 1 can be balanced out. If the ground potential at ⑥ differed from ①, the balancing for amplifier 1 would not hold for amplifier 2.

The problems encountered do not seem to relate to the shielding in the isolation transformer. A different approach seems to be required to obtain any solid benefit. Figure 6.14*b* shows an isolating transformer properly applied to one instrument. This instrument might be an oscilloscope or a tape recorder. The entire secondary of the isolation transformer is treated as an extension of the instrument's signal-shield environment and the interface between the signal shield and the power environment takes place in the isolation transformer. Now the discussion in Section 4.12 regarding double-shielded power entrances becomes applicable. The single shield within the instrument's own transformer no longer serves a useful purpose.

If the method indicated in Figure 6.14*b* is extended to more than one instrument, then in effect these instruments must have the same zero-signal reference potential. If not, the difficulties that appear in Figure 6.14*a* will prevail.

An isolation transformer is not a substitute for the proper shielding of individual instruments. If this shielding is not sufficient, corrective action can be taken but it should be taken on an individual-instrument basis. Any other arrangement may effect a change or some improvement but it will not guarantee a solution to a system-isolation problem.

The interconnection between the isolation transformer and the instrument involves shielded conductors. Since these conductors are within

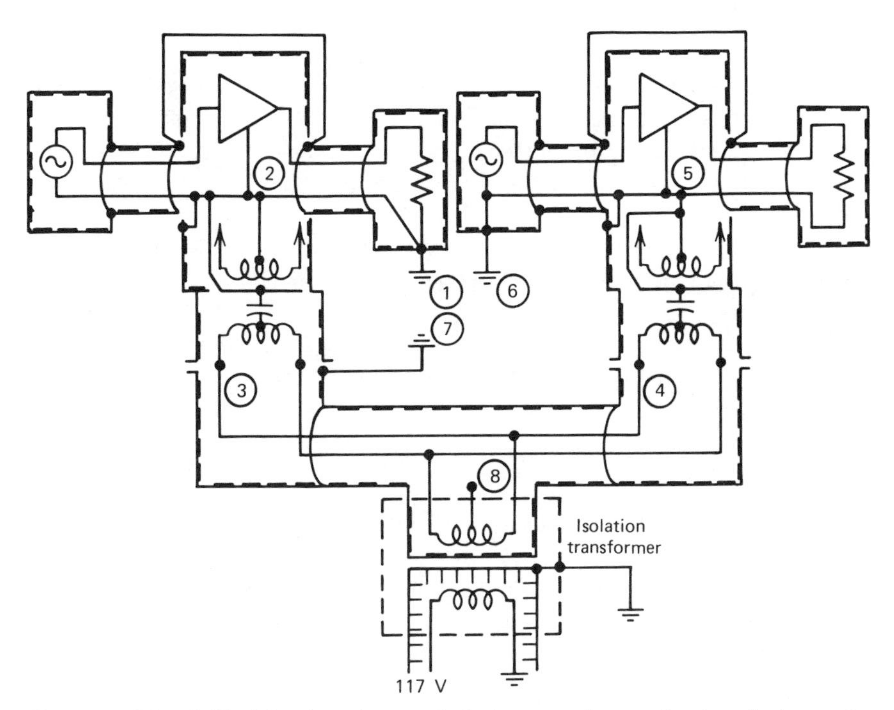

Figure 6.14*a* Difficulties in using an isolating transformer for two instruments.

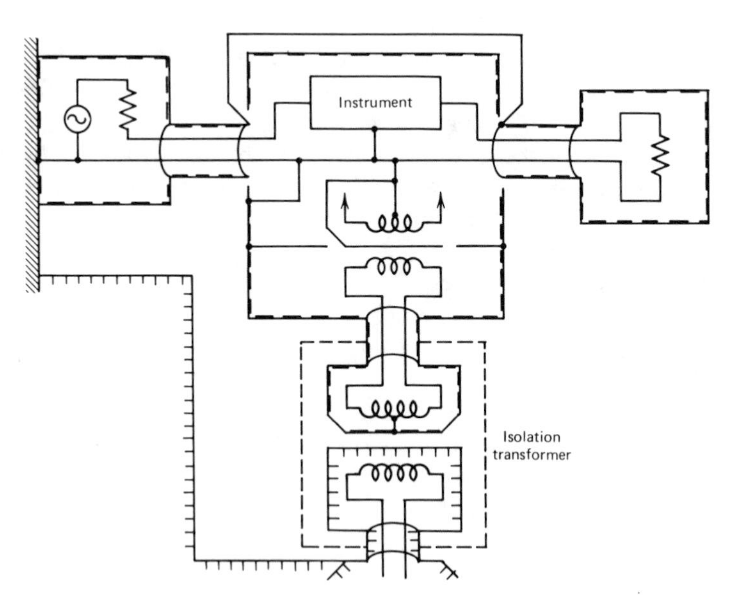

Figure 6.14*b* A properly applied isolation transformer.

the extended shield environment of the signal, they must be completely enclosed. If open wires are used, the benefits to be derived from the isolation transformer are lessened. If conduit is used, it must be treated as a shield and not be permitted to contact other grounds. This is mentioned, as isolation transformers are often remotely located for convenience.

The use of box shielding in the construction of power isolation transformers can reduce the mutual capacitance between the primary coil and the secondary shield to values below 0.001 pF. This same figure is valid for the mutual capacitance between the secondary coil and the primary shield. Shields of this tightness are academic as other mutual capacitances in a typical system cannot be held correspondingly low.

6.15 ISOLATION TRANSFORMERS FOR RACK ISOLATION

If a rack of equipment is insulated from conduit ties and from building grounds, etc., the rack acts as an outer shield for the electronics. It can also serve as the zero-signal reference potential for all output signals processed in the system. This type of rack is usually earthed in a convenient manner for safety and performance optimization. A proper rack grounding is dependent on the nature of the electronics within the rack and in some cases there may be little sensitivity to the choice of grounds.

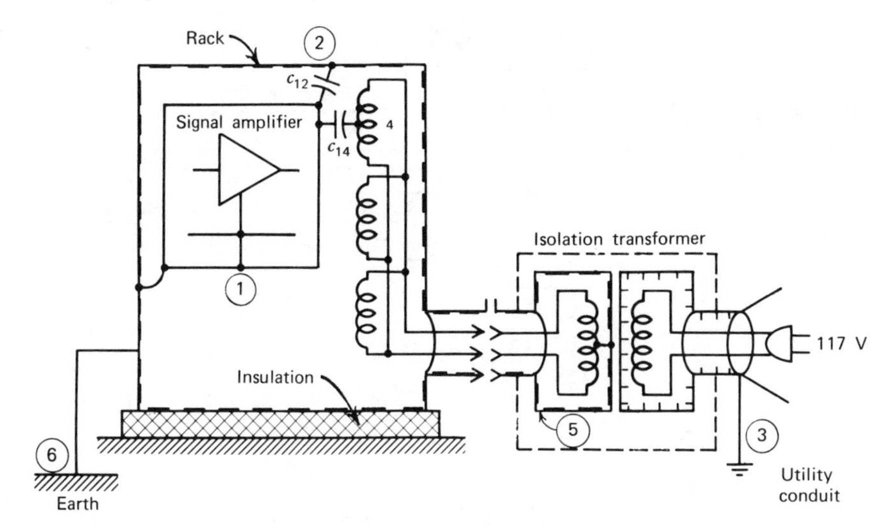

Figure 6.15 An isolation transformer used to power a rack of equipment.

Utility power can carry an unknown ground point to all transformer powered devices. To break up the mutual capacitances so that the instruments have mutual capacitances to the rack ground only and not to an unknown power ground, an isolation transformer can be used. This arrangement is shown in Figure 6.15.

Capacitance c_{12} is in parallel with capacitance c_{14}. Both of these capacitances are screened by the rack potential, which is the secondary shield ⑤ of the isolation transformer. A capacitance such as c_{13} would reduce to the mutual capacitance of the isolation transformer, which can be held to 0.001 pF if required.

The essential benefit derived by using this isolation transformer is the control of shield currents. Potential difference V_{36} causes currents to flow in shield-to-shield capacitance c_{35} in the loop ③→⑤→⑥→③. This current does not enter the signal-shield system. In effect, the double shield in the isolation transformer performs the same function for a rack of equipment that it performs in an instrument amplifier.

6.16 SINGLE-ENDED TO DIFFERENTIAL CONVERSION BY USING BUFFERS

With care, single-ended instruments can be buffered by a differential amplifier to provide reasonable differential-amplifier performance. This is often required if single-ended instrumentation is available and it must be applied to a two-ground situation. The process is not free from

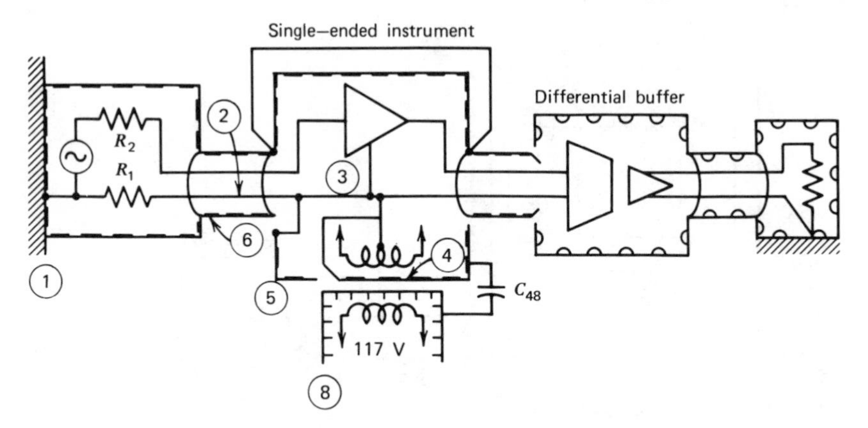

Figure 6.16 A single-ended amplifier with differential buffer.

difficulty, as is shown by Figure 6.16. A problem arises because the single-ended instrument is not intended to operate in the presence of a line resistance R_1. Capacitance c_{48} permits currents to follow the path ①→②→③→④→⑧→① if a potential difference V_{18} is present. This current flows in R_1 with the result that the amplifier sees this potential drop as signal.

Improved operation can be provided if shield ⑧ can be returned via the input shield ⑥ to ground ①. This eliminates the effect of potential difference V_{18}. Also, if R_1 is kept very low the pickup effects are lessened. It should be noted that if R_1 is 1 Ω, the currents must be below 10 μA to keep the pickup below 10 μV. It should be apparent that instruments without a second transformer shield will permit larger currents to circulate, as no conductor ⑧ exists for return to the source potential at ground ①.

The difficulty in isolating the single-ended instrument does not reside in the buffer. The single-ended instrument itself limits the performance because of its construction and design. In the extreme case, a double-shielded isolation transformer can always be added, as in Section 6.14, to provide the equivalent of shield ⑧ in Figure 6.16. Even if this procedure is followed, the full shielding precautions for the input amplifier shown in Figure 5.6 are not followed. This simply requires that the value of R_1 be kept small, certainly an acceptable limitation in many applications.

6.17 A CALIBRATION PROBLEM

The parallel connection of a group of single-ended amplifiers for calibration purposes usually creates multiple ground loops. This practice

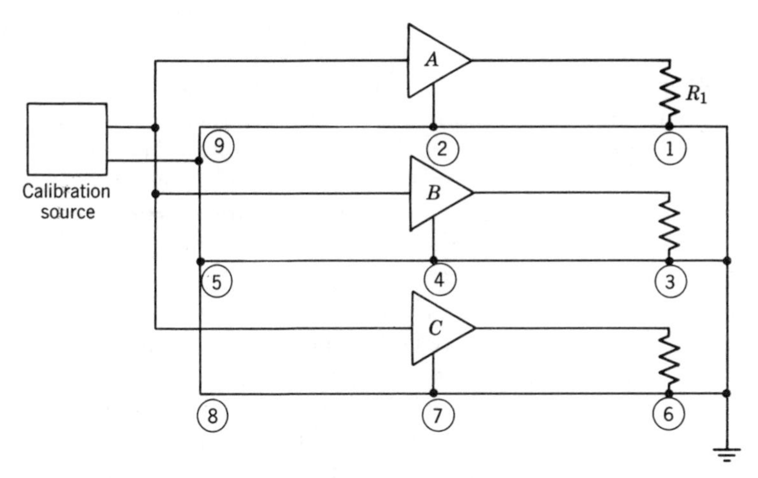

Figure 6.17 A parallel calibration scheme (*not* recommended).

is commonplace in industry in spite of this serious problem. The nature of these difficulties is emphasized so that the engineer will approach this form of calibration with caution.

Consider a group of single-ended amplifiers being calibrated from a single source, as shown in Figure 6.17. Load currents in R_1 have two alternate return paths to amplifier A other than path ①→②. These are ①→③→④→⑤→⑨→② and ①→③→⑦→⑧→⑨→②. The amount of current returning via the improper path is dependent on line resistances only. If input and output line resistances are 1 Ω, then approximately 40 percent of the load current for amplifier A will return via the alternate path using segment ⑨→②. This constitutes a significant external feedback structure, and obviously the expected gain A of the amplifier will not be realized.

The current return to amplifier A in input segment ⑨→② should cause less than the expected calibration error. If the output error is expected to be 10 mV, a 10-μV error is permitted for an amplifier gain of 1000. A 10-μV signal is picked up in 1.0 Ω for a current flow of 10 μA. Therefore the current flowing in all alternate paths must be held to below 10 μA. For a 100-mA load this means that less than 0.01% of the load current can return via all other secondary paths.

The above arguments are valid regardless of the isolation or operating qualities of the calibration source The amplifier inputs are grounded to the calibration source and to each other in the most general sense of the word grounded. Even a battery-operated calibration source will fail here.

If resistance ⑨→② is kept very small, then the effects described above can be minimized. If the amplifiers have nominal gain and bandwidth, other problems may be encountered. The multiple return-path processes can lead to system instability; that is, the amplifiers oscillate as a system while singly they are stable. The author has encountered systems where 30 channels are stable and 32 channels will oscillate. The reasons for the instability are straightforward. A stability analysis would be complicated because it would necessarily have to include the amplifiers' transfer function.

If the connections at ⑤ and ⑧ are not made, then the signal return for the calibrator is made through ③ and ⑧. This solution involves large input-signal loop areas and is not recommended. It does avoid the ground-loop problem but causes a new set of pickup difficulties.

There are two solutions to this dilemma. The first is to make amplifiers *A*, *B*, and *C* differential to accommodate the second ground connection. The second solution involves calibration-signal isolation transformers for each amplifier. These transformers should be double shielded with the shields returned to the calibration signal common and amplifier signal common respectively. The transformer precludes very low frequency or dc calibration and therefore may be undesirable.

The worst-case condition for parallel calibration exists when maximum-load currents flow at full gain. If errors are created by such a scheme they will be most apparent under these conditions. Verification of performance is best carried out by calibrating a single-channel alone and comparing this result with operation when all other channels are connected to the calibration scheme.

6.18 THE "STAR" CONNECTION–GROUND WIRING SEQUENCES IN SYSTEMS

Within one zero-signal reference system the zero reference conductor or ground must interconnect many functions. In an instrument this might include gain stages, calibration signals, load, and gain switches. These conductors should be interconnected so that potential differences resulting from signal current flow will not produce spurious feedback or pickup. This problem can be handled by either a series or "star" configuration. Both solutions provide wiring sequences so that load currents are supplied from the power supply without flowing in low-level signal conductors. In the star configuration as in Figure 6.18*a* a single point or "mecca" is used, and all loads emanate from this one point. In a series connection as in Figure 6.18*b* the loads are connected to one ground bus on the order of signal level, with the power supply being connected at the output or high level end of the bus.

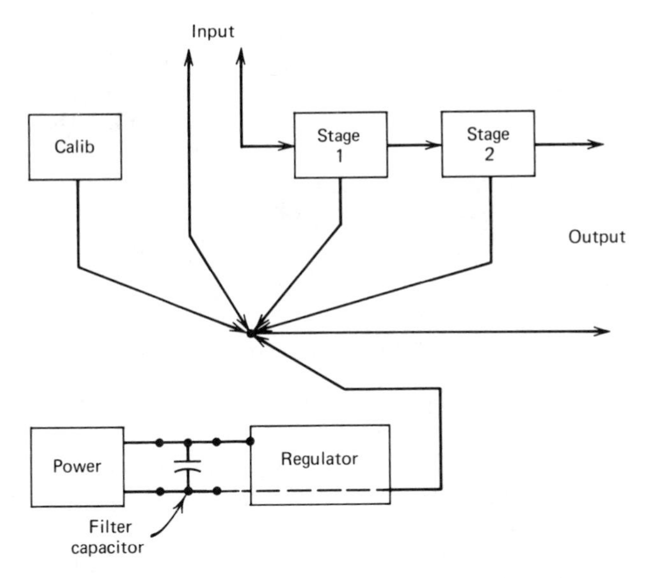

Figure 6.18a A typical star–ground configuration in an instrument.

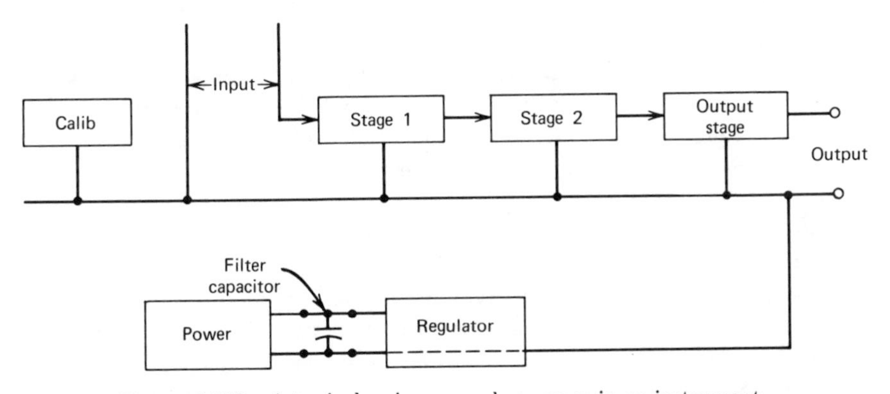

Figure 6.18b A typical series–ground sequence in an instrument.

Grounding sequences become more complex when multiple power supplies are involved. In general, the best interconnection technique keeps the load currents for each supply flowing in separate conductors. Ties or interconnections between grounds therefore handle very small currents. These connections should be made physically available for testing or change. This technique is particularly valuable when digital and analog circuits must be interconnected.

One commonly encountered current path involves power filter ripple current. This current should never share a path that involves signal processing. Capacitors should have four terminals when possible. This forces the ripple current flow to use capacitor connections other than those used to connect the rectified load to the capacitor.

When system segments must be interconnected either the star or series bus configuration can be used. Star configurations are difficult to accommodate in systems and a compromise is usually required. A knowledge of the load and signal currents can aid in any compromise.

6.19 HIGH-FREQUENCY GROUND LOOPS

The presence of an oscilloscope in an instrument system often adds a high-frequency ground loop. This loop is present even if the third-wire ground for the oscilloscope is not connected. This loop involves the capacitance between the primary and secondary coils of the power transformer (or their shields), the utility ground, and any grounds connected to the instrumentation. The oscilloscope, because of its physical size, also has capacitance to earth ground even if the utility power connection is disregarded.

High-frequency signals (impulse noise from lights, machines, or relay coils) will use any and every conductor as a reactive current path. These signals can be very difficult to trace or eliminate. One problem results because reactive current will flow in the coaxial shield of any signal probe. This current generates a signal for the oscilloscope even if the probe is shorted out at its tip. These signals are enhanced by the inductance per foot of the shield. Of course when the oscilloscope is not connected to the ground of the instrumentation, the noise will generally go away—but so will the signal.

One solution to the problem involves adding a parallel path to the oscilloscope ground from the instrumentation. This path must provide a lower impedance to the oscilloscope frame than the coaxial signal shield. If this is done, the pulse noise sensed at the probe tip will be the actual noise displayed on the crt.

It is interesting to analyze the nature of pulse type inductive signals. A relay coil of 200 mH carrying 200 mA dc and having a shunt capacitance of 20 pF produces a very high voltage when the current is interrupted. After the circuit opens, all the static energy stored in the inductance must appear across the capacitance within a $\frac{1}{4}$ cycle of the parallel resonant frequency.

Equating storage energies, we have

$$\tfrac{1}{2} CV^2 = \tfrac{1}{2} LI^2 = 4 \times 10^{-3} \text{ J}$$

or

$$V = 20{,}000 \text{ V}$$

Since this signal is sinusoidal, the maximum dV/dt is 10^{10} V/sec. Thus the peak current in each picofarad of capacitance calculates to be *10 mA*. At the natural frequencies involved here, the voltage sensed per foot[1] of shielded braid would be 5 mV. If the capacitance were 10 pF and the cable were 5 ft long, the signal sensed would be 250 mV, a large signal indeed.

The reader should note well that this is not an rf or magnetic phenomenon. It is strictly an electrostatic process. If the nature of a problem is not recognized, a useful solution will not be available.

[1] Assume the inductance per foot for typical cable to be 1 μH.

7

Shielding in Resistance-Bridge Systems

7.1 GENERAL

Resistance bridges are used extensively in instrumentation. They are used in measurements involving strain, pressure, acceleration, and temperature. The bridge may use one or more active elements or arms with the remaining fixed elements located in the instrumentation.

A resistance bridge must be excited by a power source. This may be constant current or constant voltage in nature. Batteries can be used but active power supplies are usually preferred. Active power supplies are troublesome, however, because of the shielding difficulties they present. It is the intent of this chapter to discuss the shielding required in both the power supplies and amplifiers to properly instrument resistance bridges.

7.2 THE RESISTANCE BRIDGE AND ITS SIGNAL ENVIRONMENT

All of the electronics associated ohmically with the bridge must be considered a part of the bridge-signal environment. This includes the bridge arms, its source of excitation, signal cabling, and any peripheral conditioning circuitry such as balance potentiometers or calibration elements. The parts of the system can be arranged in many different ways. The method selected in the figures below presents the fundamental ideas so that they are in their simplest form. Rearrangement of cables, etc.,

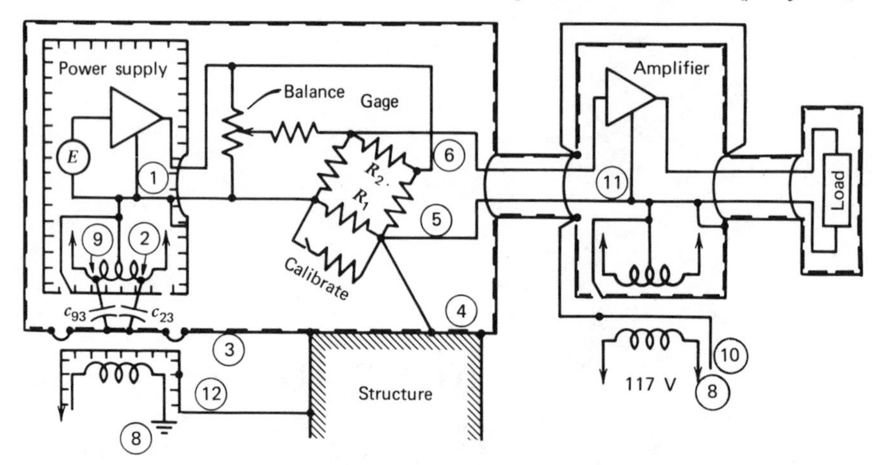

Figure 7.2 A grounded strain-gage bridge and a single-ended amplifier.

presents no logical difficulty once the essentials are understood. Figure 7.2 shows a single-ended amplifier, a strain-gage bridge, and a floating power supply, all contained in *one* signal-shield environment. Note that the gage mounting structure is included as a part of the shield closure. The power supply is correctly shown as an amplifier with a fixed input signal. The zero-signal reference conductor for the power supply and for the instrumentation amplifier are not the same. The separation of these conductors by the bridge elements provides the basis for describing the bridge excitation as a floating power supply.

7.3 THE FLOATING POWER SUPPLY

The power supply of Figure 7.2 is shown with a three-shield transformer. The inner shield is required by Rule 1, which states that a shield must surround each signal. The signals within the power supply are fixed but the shields must follow the rules for any amplifier. The inner-amplifier (power-supply) shield is connected to its zero-signal reference conductor ① at the point where conductors exit the shield enclosure.

If the inner shield is not fully effective, mutual capacitances such as c_{23} and c_{93} and coil potential differences such as V_{12} and V_{19} can circulate currents in loops such as ①→②→③→④→⑤→①, ①→②→③→④→⑤→⑥→①, ①→⑨→③→④→⑤→①, and ①→⑨→③→④→⑤→⑥→①. At first glance the effects of this current appear to be balanced out. First, the center tap of the transformer secondary can permit the mutual capacitances c_{23} and c_{93} to carry currents that are opposite in phase. Second, at the node where ⑤ connects to the bridge, the current divides so that approximately half of the current flows in a bridge arm R_1, the remaining current flowing in the adjacent bridge arm R_2 through the output impedance of the power supply. This approximate division of current into adjacent bridge arms cancels most of the signal picked up by the amplifier.

Although these balancing processes are present, the flow of current is undesirable for the following reasons:

1. Currents that flow in long-line signal conductors cause signal pickup that cannot be balanced out.
2. The unbalance in current division caused by the output impedance of the power supply develops an unwanted signal for the amplifier.
3. If bridge balance is required to eliminate pickup, then an unbalance during signal generation would permit the undesired signal to be modulated by the desired signal.
4. The ac currents that flow in the bridge resistances also flow in the parasitic capacitances associated with the bridge arms. Reactive balancing at 60 Hz is complex at best and should be avoided. If the currents are avoided, then reactive balancing is not necessary for 60-Hz rejection.

The above arguments all support the need for the inner shield in Figure 7.2. This shield reduces the flow of power currents in the bridge system so that the undesirable effects just described are minimum.

The primary shield in the instrument amplifier insures that currents will not flow from the primary coil through the signal conductors. For example, a capacitance $c_{8,11}$ would permit current to circulate in the loop ⑧→⑪→⑤→④→⑧ as a result of potential difference V_{48}. If the strain-gage system is grounded at the instrumentation output it cannot be grounded again at the bonding surface. This means that the shield closure cannot be complete at the most critical point, and the parasitic currents that flow in the bridge elements can be quite troublesome. This is a basic limitation where single-ended instrumentation is applied. A proper solution only exists if the potential at the structure can be controlled. See Section 6.6 for a further discussion of this problem.

7.4 FLOATING POWER-SUPPLY SHIELDING

The effectiveness of shields can be described in terms of mutual capacitances. Since the potentials that cause currents to flow can vary considerably, this measure alone is not sufficient to describe the problem. A more practical specification limits the unwanted current flow or defines the expected performance under actual operating conditions.

The unwanted current levels that are permitted to flow in the bridge arms can be calculated roughly as follows: For 1000-Ω bridge resistances and for a 10 percent line unbalance, a 1-μV pickup results for a current of 10 nA. For 60-Hz pickup where the coil potential is 10 V, the mutual capacitance balance should be 2 pF or less. This value assumes the bridge itself is reactively balanced. If only one bridge resistance is active the reactive unbalance that results places a further restriction on the permitted current flow. A value of current sometimes used is 2 nA. From the above arguments, the secondary or inner shield should have a mutual capacitance balance of approximately 0.4 pF. In some transformers this balance is achieved through careful construction techniques. The problem is an economic one and a conventional shield is often preferred because the production problems are easier to control.

The center or signal shield of the power supply appears to be not too critical if the primary shield is connected to zero-signal reference as shown in Figure 7.2. With this connection, the potential difference $V_{4,12}$ is zero, and no 60-Hz current will circulate through mutual capacitance $c_{1,12}$ in the loop ①→⑫→③→④→①. When the primary shield is not present the mutual capacitance from ① through the center shield terminates on conductor ⑧. The potential difference V_{48} can then circulate current in the loops ①→⑧→④→① and ①→⑧→④→⑥→①. By previous arguments, this current should be below 2 nA. The combined effectiveness of the signal and primary shield for current flow which involves conductor ⑧ should be below 0.1 pF to allow for primary voltages that average 50 V. (The 0.4 pF figure was based on a 10-V assumption.)

If the signal shield were omitted and the primary shield alone provided the mutual capacitance calculated above, the electrostatic closure of the signal would be technically correct in a segmented sense. This procedure, because of cable lengths and high-frequency noise pickup, is not optimum. The circuit approach is shown in Figure 7.2 is preferred.

7.5 MULTICONDUCTOR CABLE

Typical bridge connections can involve up to eight conductors. This number is used to provide separate connections for bridge-voltage metering and calibration. Without these extra conductors calibration errors are easily created because of line-resistance effects. Also, without these conductors, bridge metering is in error because of the potential drop in the bridge supply lines.

These conductors can be in one shield enclosure if desired. In some installations, a separate cable is run for calibration and metering. The shields must still be connected once to signal-zero reference. Multiple connections to ground are in violation of the basic shielding rules.

7.6 DIFFERENTIAL AMPLIFIERS AND RESISTANCE BRIDGES

Differential amplifiers can be advantageously used in resistance-bridge instrumentation. The two signal-zero reference potentials permit applications not available with single-ended instruments. Figure 7.6*a* shows a resistance bridge in a typical application. The input shield must be connected to the signal-zero at one point only. If this point is selected as the power-supply return conductor, the input signal shield serves to enclose the power-supply circuitry. This connection is made from point ① to ②.

The input signal can be grounded once. The obvious ground point to use is the structure being measured. With this connection, the structure provides the proper potential to enclose the signal at the bridge. With the through connection from ① to ③, the power supply is grounded. The stringent floating requirements of Section 7.4 are thus not required.

The power entrances for this application are shown in Figure 7.6*b*. Two electrostatic shields are indicated for the power supply. This insures that the potentials on coil ① will not circulate currents in c_{12} through the loop ①→②→③→④→①. The second shield in the amplifier insures that currents will not circulate in the output-signal conductor.

Because the potential drops in conductor ③ produce common mode for the amplifier and are rejected, and because currents flowing in the amplifier-output conductor are not subject to amplification, the flow of unwanted current appears to be noncritical. The factors requiring the

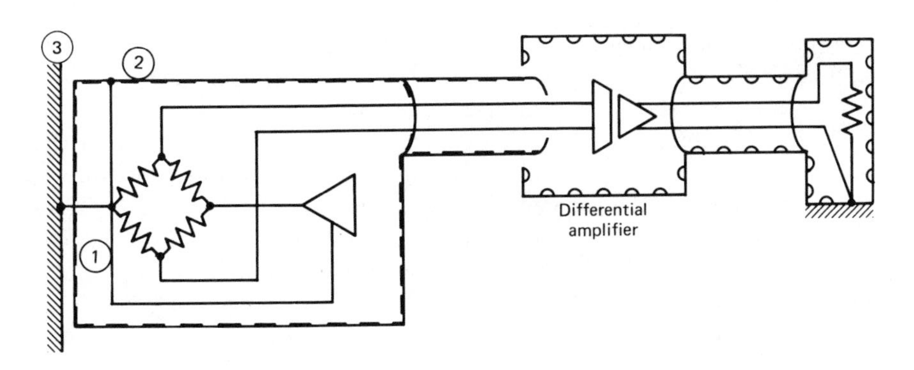

Figure 7.6*a* A differential amplifier and a resistance bridge.

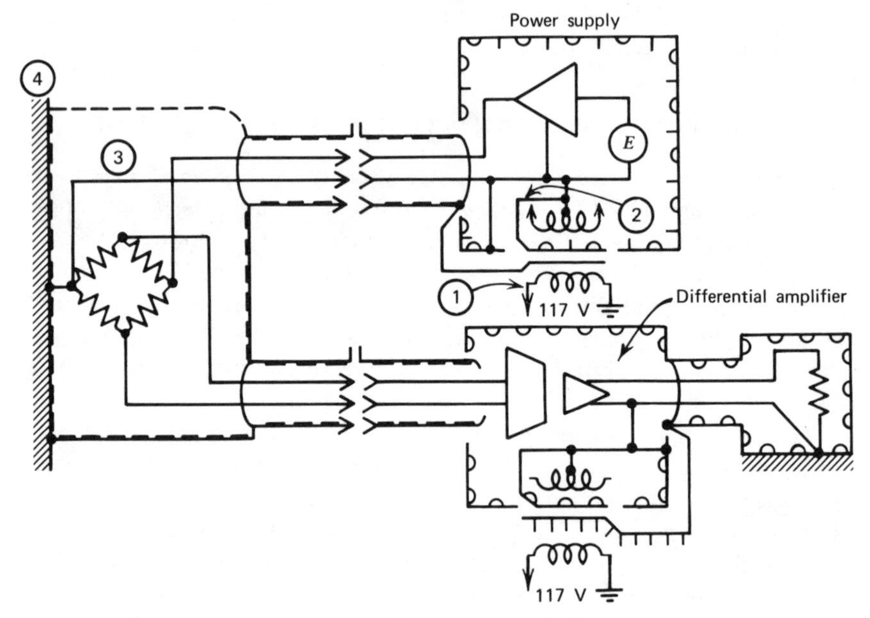

Figure 7.6b Power entrances for differential amplifier and resistance bridge.

extra shielding involve some secondary considerations. For example, common-mode signals are not easily rejected if they are high frequency in nature. This type of pickup is apt to occur if the primary shielding is not utilized.

7.7 COMMON POWER-SUPPLY EXCITATION

Groups of resistance bridges are often applied to a measurement. The approach shown in Figure 7.6*b* suggests that a common power supply can be used to excite many bridges at one time. This is indeed possible as long as the bridges are all grounded to one structure potential. If the structure has a potential gradient on its surface, then this procedure is not recommended. The problem that ensues is shown in Figure 7.7*a*. Here the potential V_{12} and a typical capacitance c_{23} permit a loop current ①→②→③→④→① to flow. Since this current flows in the bridge resistances, the resulting signal pickup is amplified. If only a part of the bridge is active, then the reactive unbalance increases the likelihood of unwanted pickup. If the potential gradient is negligible, the common power supply can be used.

In applications where a cluster of resistance bridges is used in a

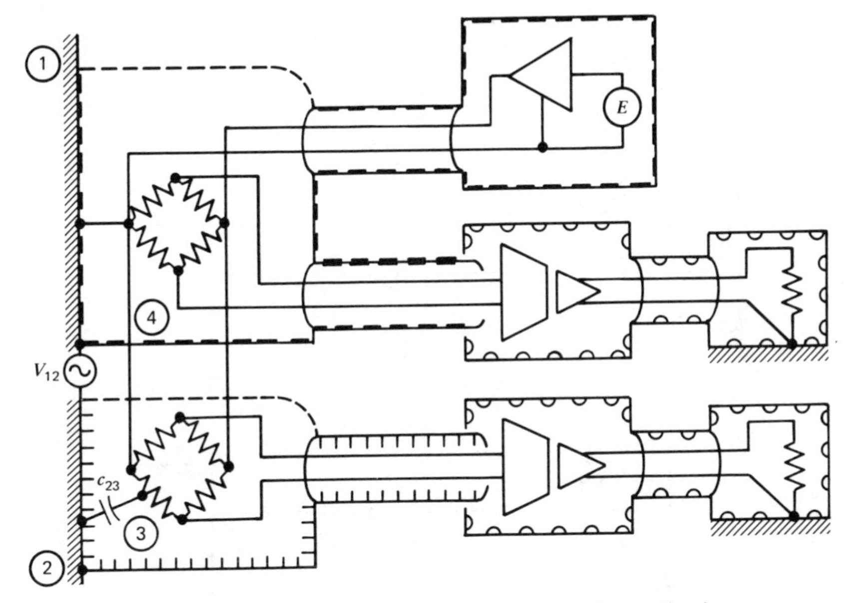

Figure 7.7*a* An improper common power-supply application.

single potential environment, a common power supply is recommended. This arrangement is shown schematically in Figure 7.7*b*.

7.8 SHIELDING CALIBRATE PROCESSES

Resistors can be shunted across the arms of a resistance bridge to simulate resistance change. This calibration procedure takes on many forms and can be manually switched or remotely controlled by relays. These relays and their associated coils must interface the signal processes and therefore they must be properly shielded to do an effective job. Shielded relay structures are not as readily available as shielded transformers, but since both can contaminate the signal, both require attention. Figure 7.8 shows a properly shielded relay applied to a calibrate resistor. If the shield is not placed between the coil and the resistance bridge, the mutual capacitance c_{12} or c_{13} can provide a noise source by permitting currents to flow in the loop ①→②→④→①.

Relays are normally energized by a switch closure. This places a high-voltage step function on the coil that can easily couple to the signal. The shielding in Figure 7.8 reduces this influence. The relay shield protects against the relay-power ground potential, which is not generally at the signal-zero reference potential.

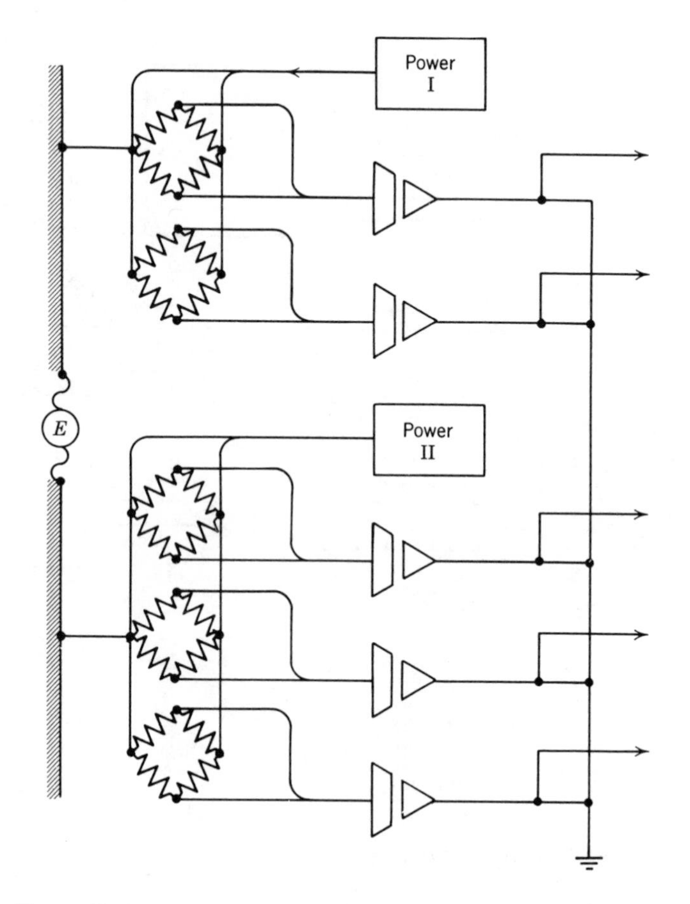

Figure 7.7*b* Separate power supplies of groups of resistance bridges.

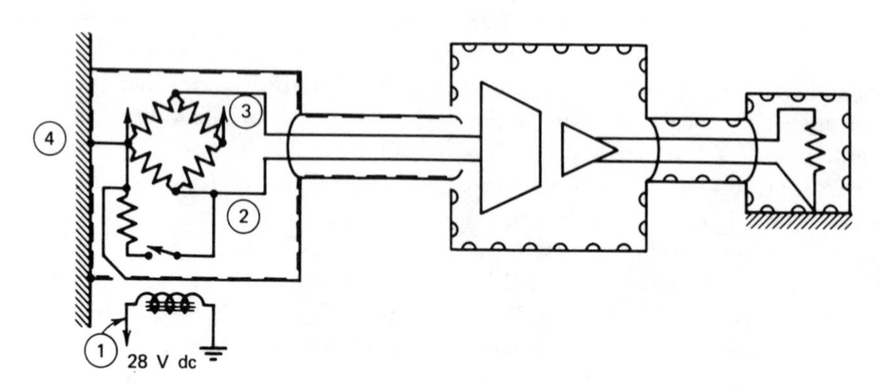

Figure 7.8 A shielded relay for calibration.

The shield treatment in Figure 7.8 is electrostatic only and does not prevent magnetic coupling. This noise problem is best treated by wiring signal circuitry in a minimum loop-area configuration. This problem is treated in more detail in Chapter 8.

7.9 AMPLIFIER POWER-SUPPLY COMBINATIONS

Figure 7.6*b* shows a differential amplifier and a separate power supply processing one resistance bridge. The signal-zero reference for the power supply and amplifier input are at the input structure. The output of the amplifier is grounded separately to the point of observation. If the power supply ground and the amplifier-output ground are shared in one instrument, the ground connection at the structure is no longer possible. This invites parasitic difficulties as previously discussed. Although this arrangement is more economical (one power transformer) it is not recommended where low-level signals are to be amplified.

8

Magnetic Processes in Instrumentation

8.1 INTRODUCTION

The intent of this chapter is to discuss the magnetic field in specific areas only. A discussion in depth is outside of the scope of this chapter and would not result in an improved understanding of the instrumentation problem.

The transformers used in previous discussions required magnetic fields for transformer action. These components were used in both power entrances and in signal coupling. Although transformer design is of engineering importance, here again many facets of this art will not be discussed for the reasons given above.

Magnetic fields are found where cables carry current and when utility power is distributed. The electromagnetic field is also present wherever transmission-line phenomena are present. It is these general areas that will receive attention.

8.2 BASIC IDEAS

A magnetic field can be described in terms of lines of force or tubes of force just as in the electrostatic case. Where the field E due to charge distribution was measured by a unit charge, in magnetics the unit magnetic pole is used to measure the magnetic field. This idea can be used to develop the subject of magnetic field theory through the use of vector

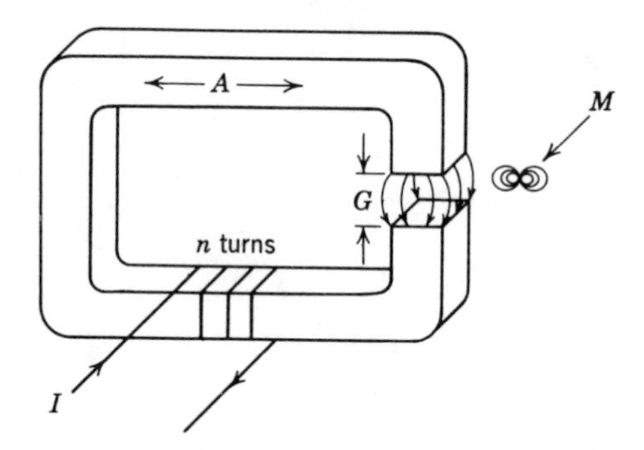

Figure 8.2a A gapped case with a magnetizing current.

potentials. Discussion based on this approach can be found in most standard physics texts. A more pragmatic approach will suffice in this chapter.

Magnetic fields are present when charges are moved. If current flows in neighboring conductors, forces exist that tend to modify the conductor geometry. These forces are described in terms of the magnetic field surrounding the conductors. (These forces tend to rearrange the conductor geometry so that the system stores less magnetic field energy.)

Magnetic materials can provide a source of static magnetic field without current flow. Since these fields are not likely to involve instrumentation-system processes they will not be discussed here.

The behavior of the magnetic field is dependent on the conductor geometry and upon the materials near the conductors. If these materials are iron, for example, the fields tend to concentrate in these materials indicating that they prefer this simplest path. In a more accurate sense, the energy stored in the system is minimum when the field follows the path of the magnetic material.

A parallel in electrostatics can easily be found. The displacement field D is determined by charge Q only. The force field E is dependent on the dielectric materials present. The energy stored is proportional to this field E. To minimize potential energy, the field E tends to concentrate in regions having a high dielectric constant.

A magnetic field discussion can start by considering a simple magnetic circuit. In Figure 8.2a a current I establishes a magnetic field in the ferromagnetic path A. This path is broken at one point by a gap G. If a small permanent magnet (a unit magnetic pole) is placed into this gap, the force on this magnet can be used to measure the magnetic flux density in the gap. This flux density is given the symbol B. If experiments are conducted, the following relationships are found:

1. The field B in the gap is proportional to the product of current and turns, or nI.

2. The field B in the gap is inversely proportional to the gap length for constant current.

3. If the gap is doubled, the current must be roughly doubled to obtain the same field.

These conditions are analogous to the current flow in Figure 8.2*b*. The battery V is analogous to the ampere-turns nI developing the field. The current flow is analogous to field B. The resistance R is analogous to the gap. The conductors carrying the potential are analogous to the magnetic path carrying the magnetic field to the gap, and the resistance of the conductors is analogous to the limitations of the magnetic material.

The gap controls the flux density B in the same way the resistor controls the loop current in Figure 8.2*b*. Because the field follows the magnetic material to minimize system energy, it is proper to say that most of the energy is stored in the gap and very little in the magnetic path. The circuit analog shows the same thing, namely that energy is dissipated in the resistor and very little in the conductors connecting the resistor to the voltage source.

The magnetic circuit is involved with the path length of the magnetic field whereas the circuit idea involves discrete elements. It is this difference that must be considered next. The experiment in Figure 8.2*a* indicated that the ampere-turns had to increase with gap dimension if the flux density B was to be constant. It can also be shown that even the best magnetic path is not perfect; thus increasing its length also requires increased nI. This simply demonstrates that ampere-turns, magnetic path length, and the nature of the magnetic material are all closely related. In particular, the ampere-turns required to establish the field B are proportional to path length in every segment of the magnetic loop. The gap requires the largest number of ampere-turns, the magnetic material, the least.

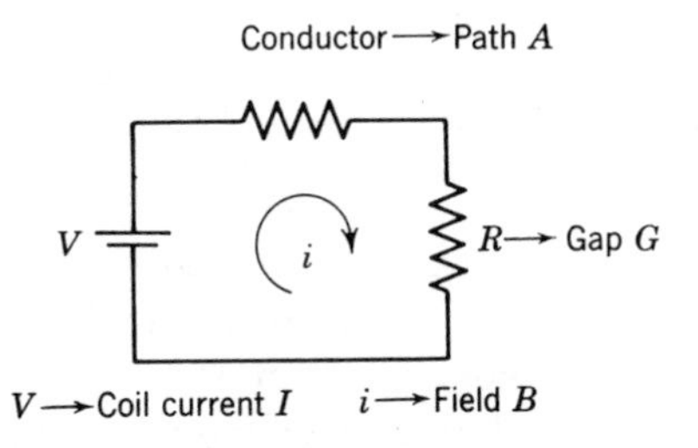

Figure 8.2*b* A simple circuit analogy to Figure 8.2*a*.

The ease of magnetization is measured as a ratio between the flux density B and the ampere-turns per unit length required to establish the field B. This ratio is defined as the permeability of the material. If the symbol H is used for the magnetizing force in ampere-turns per meter, the permeability of a substance is just the ratio

$$\mu = \frac{B}{H} \tag{1}$$

In the magnetic circuit of Figure 8.2*a* the magnetizing current required to establish a field B can be derived as follows: In the magnetic path A, the field B requires an H_A value of B/μ_A where μ_A is the permeability in path A. In the gap G the H_G value is B/H_G where μ_G is the permeability of the gap G. The ampere-turns required in any path length is simply the product of H times the path length l. For the magnetic material in path A the ampere-turn requirement is $H_A l_A$. Similarly in the gap, the ampere-turn requirement is $H_G l_G$. The ampere-turns required for a flux density B in the total magnetic path becomes

$$nI = \frac{B}{\mu_A} l_A + \frac{B}{\mu_G} l_G \tag{2}$$

For the case where the gap is zero, the magnetic-flux density is simply

$$B = \frac{\mu_A nI}{l_A} \tag{3}$$

Equation (3) illustrates that the field B is proportional to permeability as well as to the current.

8.3 LENZ'S LAW

Early experimenters in magnetic processes determined that currents could be made to flow in conductors by moving them past a magnetic field. It was also found that a changing magnetic field and a stationary conductor produced this same phenomena. Experimentation revealed the following facts.

1. The current was dependent on the loop area of the conductors.
2. The current was greater if the magnetic field changed more rapidly.
3. If a multiturn coil was used, then the voltage appearing at the coil ends was proportional to the loop area A, the turns n, and the rate of changing flux.

These facts can be expressed mathematically as Lenz's law

$$V = -nA\frac{dB}{dt} \tag{1}$$

The minus sign is used to show that if current is allowed to flow in the coil, the magnetic flux established by this current would oppose the flux linking the coil.

The geometry of the field B can be represented by lines of flux. As in the case of the field E, these lines can be assigned on any convenient basis to establish the contour of the field. The field B is completely described in terms of line density and the directions these lines take. If the flux itself is of interest, it is simply the product of flux density times the area normal to the lines. Equation (1) expressed in terms of flux becomes

$$V = -n\frac{d\phi}{dt} \tag{2}$$

Here ϕ is the flux crossing the turns n and is equal to the product BA.

Equation (1) relates voltage to field strength B and does *not* contain a permeability factor. The B field is therefore directly responsible for inducing voltages. For this reason B is also titled the field of magnetic induction as well as flux density.

8.4 AMPERE'S LAW

The ideas presented in Section 8.2 can be extended to provide a valuable tool. It was shown that the sum of products of $H \cdot l$ around a magnetic loop equals the ampere-turns threading that loop. Expressed in integral form,

$$\int H \cdot dl = nI \tag{1}$$

This equation is also called Ampere's second law.

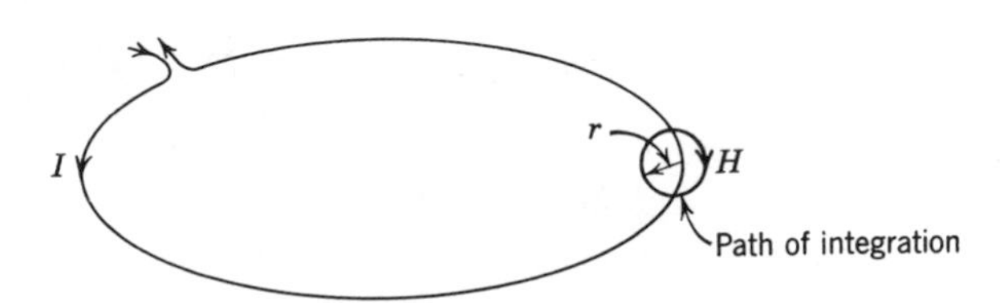

Figure 8.4 The magnetic field near a wire.

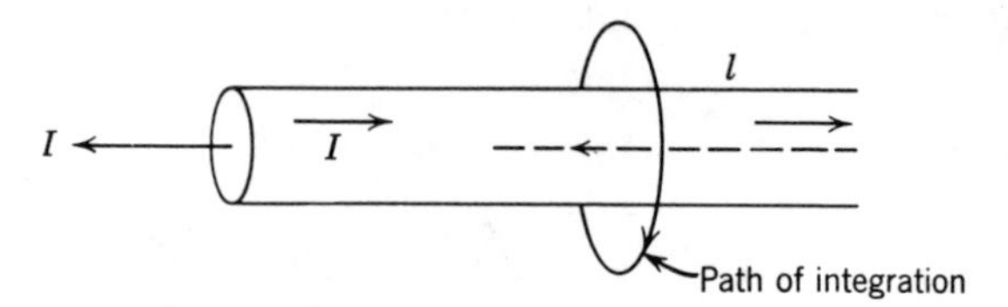

Figure 8.5 Coaxial current flow.

If a very large loop of current is considered this summation can be carried out near the conductor. (A large loop is selected to insure that the principal field is caused by the nearby current and a small contribution from the far side.) The field H at a distance r from the conductor is constant, and a direct application of Eq. 8.4(1) in Figure 8.4 yields

$$\int H \cdot dl = 2\pi r H = I \tag{2}$$

or

$$H = \frac{I}{2\pi r} \tag{3}$$

8.5 COAXIAL CURRENT FLOW

If current flows in a coaxial manner as in Figure 8.5, the external magnetic field is zero. This results because equal and opposite currents thread the loop l. This is obvious by direct application of Eq. 8.4(3). (The complete absence of external field does not actually occur as this would require that the two currents be accurately coaxial and that the currents flow in an infinite straight line.)

Parallel conductors carrying equal and opposite currents will have a small external field. Symmetry does not exist near the conductors, but at large distances this fact is of little importance. Parallel conductors are often twisted to maintain proximity and reduce any residual loop area effects to reduce the external field even further.

8.6 MAGNETIC LOOP AREAS

The discussions in Sections 8.3, 8.4, and 8.5 demonstrate the important fact that loop area plays a dominant role in magnetic behavior. If magnetic flux crosses a loop a voltage appears in that loop. Conversely, if current flows in a loop, a magnetic field is created. It is thus axiomatic to state the following:

1. All critical signal wiring should have minimum loop area to avoid pickup from external magnetic fields.

2. Currents should flow in minimum loop areas to avoid magnetic field generation.

8.7 MAGNETIC UNITS

The units used in magnetics can vary depending on the problem area, the particular engineering discipline involved, and the ease of calculation. The unit of magnetic induction B is defined by the torque on a small loop of current in the field. If the product of loop current and loop area IA is unity, this is defined as a unit magnetic moment. If a unit magnetic moment is placed in a field of 1 weber per square meter ($\mathrm{Wb/m^2}$), it causes a torque of 1 newton-meter (N-m). The field B can also be measured in gauss. One G is equal to 10^{-4} $\mathrm{Wb/m^2}$. One line/$\mathrm{cm^2}$ is equal to 1 G. A line is also called a maxwell (Mx).

When current is equated to the field of magnetic induction in a vacuum, the permeability becomes involved as a factor. Equation 8.4(3) can be expressed in terms of the field, i.e.,

$$B = \frac{\mu_0 I}{2\pi r} \tag{1}$$

This expression can be used to define μ_0 or I. If μ_0 is given the value $4\pi \times 10^{-7}$ Wb/amp-m, then (1) experimentally defines the ampere. Note that this same permeability factor relates B and H in a vacuum.

In a medium other than a vacuum the permeability factor μ must be introduced, and (1) can be written

$$B = \frac{\mu \mu_0 I}{2\pi r} \tag{2}$$

If B is expressed in gauss, I in amperes, r in centimeters, then, setting $\mu_0 = 4\pi \times 10^{-7}$, (2) becomes

$$\boxed{B = \frac{0.2\mu I}{r}} \tag{3}$$

where μ is the conventional dimensionless permeability factor.

Lenz's law, Eq. 8.2(3), states that voltage is proportional to the time rate of change of flux. Expressed in convenient units, the induced voltage becomes

$$\boxed{V = -10^{-8} nA \frac{dB}{dt}} \tag{4}$$

where V is in volts, A is the loop area in cm^2, and B is in gauss.

8.8 MUTUAL AND SELF-INDUCTANCE

When a changing current causes a changing magnetic field to cross the conductors carrying the current, a voltage is induced with the conductors by Eq. 8.7(4). This induced voltage is dependent on geometry, turns, and the permeability of the material in the geometry. Since any flux ϕ is dependent to the current flow I, 8.7(4) can be written as

$$V = -10^{-8} nk \frac{dI}{dt} \tag{1}$$

The factors in (1) can be lumped together as one factor called the self-inductance. Thus (1) becomes

$$V = -L \frac{dI}{dt} \tag{2}$$

where L has units of henries. One H of inductance is defined as that geometric property where a 1 amp/sec change produces a back emf of 1 V.

Mutual inductance simply relates the voltage induced by flux coupling between circuits. If a change of current in circuit 1 produces a voltage in circuit 2, the ratio of these quantities is defined as mutual inductance. Expressed mathematically:

$$V_1 = -M_{12} \frac{dI_2}{dt} \tag{3}$$

Mutual inductance is also expressed in henries. It can be shown that in any two circuits, the mutual inductance as measured from either circuit is the same, that is

$$M_{12} = M_{21}, \qquad M_{13} = M_{31} \quad \text{etc.} \tag{4}$$

8.9 SIGNAL CIRCUIT COUPLING BY MAGNETIC FIELD

Low-level signals can be contaminated by small amounts of magnetic pickup. To indicate the problem consider a loop area of 1 cm^2 at a signal interface. If the 60-Hz field is 10 G, the pickup, by 8.7(4), is

$$V = 10^{-8} nA \frac{dB}{dt} \tag{1}$$

Since $B = 10 \sin 2\pi \cdot 60t$, $dB/dt = 1200\pi \cos 2\pi \cdot 60t$ and $V = 37.8$ peak μV.

Moderate magnetic fields, 10 G or greater, can be found near fans, power transformers, circuit breakers, and so on. It is thus important to keep loop areas minimum in connectors and cable entrances. These are likely places to pick up unwanted signals from nearby magnetically operated devices.

8.10 AN ELECTROSTATIC-SHIELD PROBLEM

Consider an electrostatic shield placed around a coil in a transformer. This shield acts as a single turn and therefore there is a potential difference at the shield ends. If the measurement is made in a special way the voltage can appear to vanish. This principle is shown in Figure 8.10. In the *A* measurement, the voltmeter leads recircle the magnetic flux so that no net area is presented for flux coupling.

This point must be considered if the ultimate in shield effectiveness is discussed. The voltages on the shield can cause current to flow in the distributed capacitances. Each element of capacitance and its current loop has a flux-capture cross section. This flux capture causes a voltage in series with the element of capacitance.[1] The problem of calculating the currents in the simplest example would be difficult indeed. This cur-

[1] The flux capture by a shield can be a serious problem if the flux is that of the leakage field. In power transformer applications in which capacitor input filters are used the leakage field can undergo abrupt changes that will result in very potent pulses of current in the mutual capacitances terminating on the shields. If these pulses of current flow in external circuits, they can easily cause voltage spikes of very short length to appear on signal conductors.

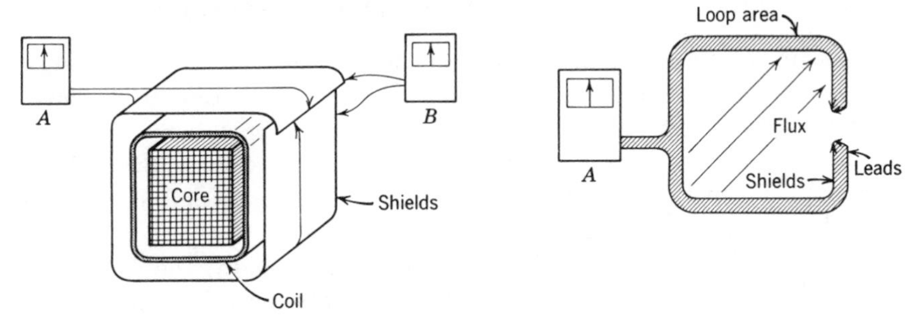

Figure 8.10 Two methods of voltage measurement on a single shield loop.

rent flow can theoretically be balanced out by proper placement of the shield contact. This effect is second-order except where the number of turns is small, that is, in a low voltage-distribution arrangement.

Very tight transformer shields can be achieved, that is, mutual capacitance of below 0.001 pF. These capacitances are measured, however, with the transformer out of operation. Since the shield can couple signals as a single turn, the currents that flow can be larger than the 0.001 pF would indicate. The true effectiveness of a tight shield should be made through a measure of operating-current flow (this current has been called by some an insertion current).

8.11 PARALLEL CABLE RUNS

Long parallel lines containing signals and power provide an opportunity for magnetic interference. If the signal circuits have small loop areas, the pickup in each signal path will be minimum. The area between adjacent signal conductors does, however, provide an opportunity for flux coupling. This flux causes a potential difference between zero-signal references. If the practices outlined in earlier chapters are followed, the added pickup will be minimum. It is important to realize, however, that the electrostatic shield is bypassed by the magnetic field. If the flux couples into a loop of signal conductors, the potential that is developed between these signal conductors and the shielding is immaterial. If these potentials cause current flow in any source impedance, the effect is obvious.

Best practice thus dictates that all power conductors, control lines, and even ground wires carrying current should be grouped together and run in a separate bundle. This insures that any magnetic coupling will be minimum. The use of shielded or coaxial signal conductors is not sufficient to ward off problems. It is safer to remove the source of the difficulty.

8.12 FLUX COUPLING TO SHIELD CONNECTIONS

In many of the instrument configurations shown in earlier chapters, a primary shield conductor is shown tied back separately to a source of zero-signal reference potential. If this conductor follows the path of the signal leads, the loop area involving signal conductors will be minimum. Since this shield conductor carries no significant current, it in itself is not a source of magnetic field.

If the shield conductor takes a second return path, the loop area that is formed is necessarily larger. The flux entering this loop modifies the shield potential at the transformer and this can result in current flow in the signal conductors. This is a second-order effect in most cases, but since it is simple to adhere to a correct practice, it is recommended here that the shield conductor be returned with the signal bundle.

8.13 USE OF CONDUIT FOR FIELD REDUCTION

The magnetic field around a single current-carrying conductor is given by 8.7(3). This field is dependent on the permeability of the material at the radius of interest. If conduit is placed around the conductor the field outside the conduit is the same as before the conduit was added.

A magnetic enclosure can be effective in reducing an external magnetic field when the entire current loop is within the enclosure. This cannot be true for a single conductor or for a circuit when the earth is used as a return path.

8.14 PRACTICAL TRANSFORMER SHIELDS

Transformers with special shields are usually made to order and are available from most transformer manufacturers at additional cost. Isolation transformers are an exception as many sizes are available as standard items.

The ideas of shielding are simple enough, but in actual practice, effective and economical techniques are sometimes closely guarded secrets. Several obvious approaches are available and these are discussed below.

1. Shielding by separation. Here the coils are separated by placing them on opposite legs of a core. The disadvantages are:
   a. The core acts as a coupling element unless it is connected to one of the coils or to ground.
   b. Leakage reactance is apt to be high.

2. Faraday shields. In this technique a layer of copper or aluminum foil is laid between coils. Further insulation permits a second shield to cover the first if required. See Figure 8.14. The disadvantage is the following:
   a. This shield is only moderately effective. Mutual capacitances to points outside of the shield can be 50 to 100 pF.

3. Box shields. The wraps of copper or aluminum are folded to box in the sides of each coil. This method provides a very tight shield. The disadvantages are:
   a. Cost. This procedure is usually carried out by hand labor.
   b. The overlap must be carefully prepared to avoid the shorted turn. The pressure of a second coil must be considered.
   c. The shields are only effective if the lead dress is carefully treated.

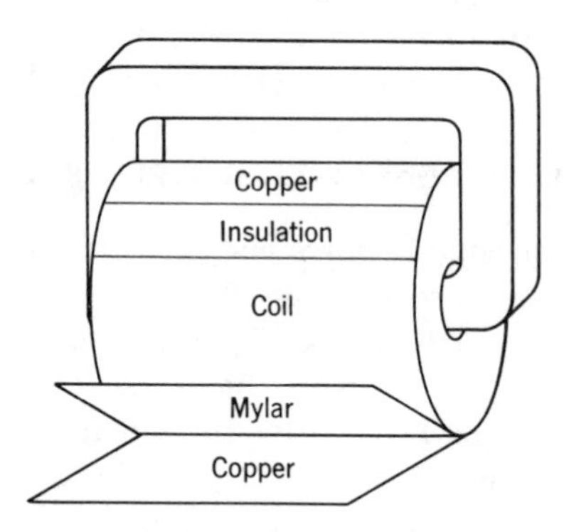

Figure 8.14 A simple Faraday shield.

These leads must also be shielded to keep the mutual-capacitance terms low.

d. Interweaving coils become extremely difficult to handle. Box shields and simple transformer geometries go together.

Some manufacturers prefer to construct the shielded coils separately. This way the coils can be nested one inside the other or side by side. Since the coils are separate entities, the pressure of one coil on another is avoided.

Shields tend to increase the effective input capacitance and leakage inductance of a transformer. These factors are in the direction to restrict transformer bandwidth. These problems are usually left for the transformer designer but they should be understood by the engineer making demands in a transformer design.

Shields can be made of conductive paper as well as nonferrous metal foil. The resistance per square of conductive paper permits shorted turns without ill effect. The point-to-point resistance is quite small compared to any leakage reactances involved in the transformer. This means that there are no potential gradients along the shield caused by reactive current flow. A paper shield has the advantage of being thin and inexpensive. Adhesive techniques are available to make connections to this type of shield.

8.15 ULTRASHIELDED ISOLATION TRANSFORMERS

The techniques of box shielding the individual windings of a transformer are simple in concept but often difficult in practice. Since the transformer is the place where the primary and secondary circuits come closest together, it is absolutely imperative that the shielding be as complete as possible. This means:

1. No pinholes in the shields.

2. Primary and secondary lead wires be shielded so that electrostatic coupling cannot take place between the leads.
3. No open spaces around leads where they emerge through the shields.
4. Excellent insulation so that leakage currents between coils are minimized.
5. Avoidance of shorted turns.

By carefully following proper techniques, effective interwinding capacitances less than .05 pF ($.05 \times 10^{-12}$ F) and leakage resistances greater than 10,000 MΩ ($10{,}000 \times 10^{6}$ Ω) can be guaranteed. Actually, values less than .001 pF and greater than 100,000 MΩ can often be obtained, but with more difficulty.

The box-shielding technique has been successfully applied to the coils of wire that enclose the following magnetic cores:

1. E-I laminations
2. C cores
3. Toroids
4. Cup cores
5. 3-phase E cores

The following figures show some of these techniques in practice:

Figure 8.15a Large transformer coil before shielding (left) and after shielding (right). Note E lamination which will be used with this coil. (Photos courtesy of Topaz, Inc., San Diego, Calif.)

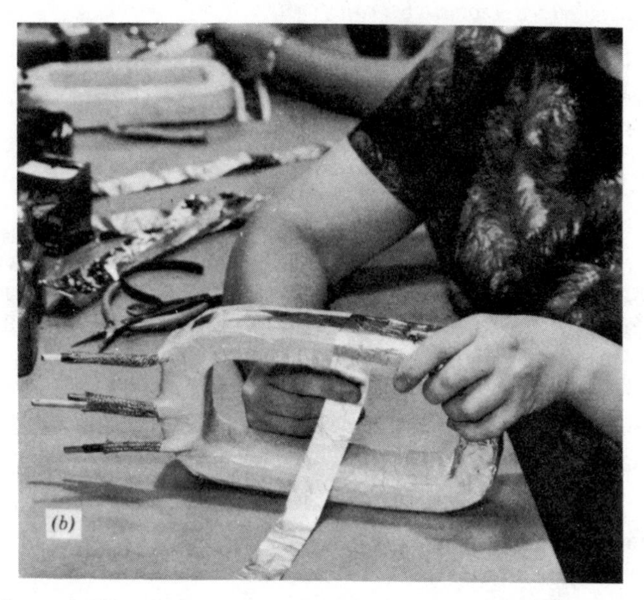

Figure 8.15*b* Coil of Figure 8.15*a* receiving a second box shield.

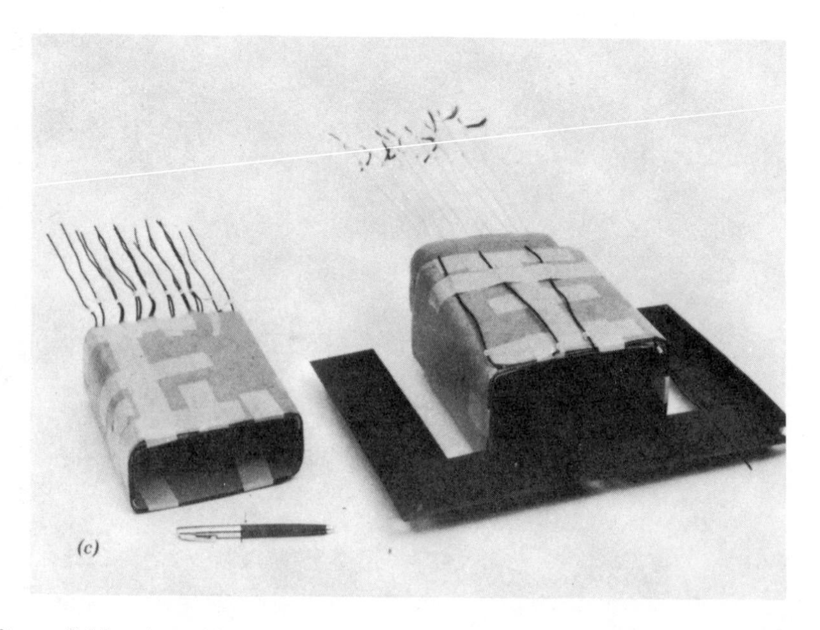

Figure 8.15c Assembly of transformer with inner coil box-shielded and outer coil unshielded.

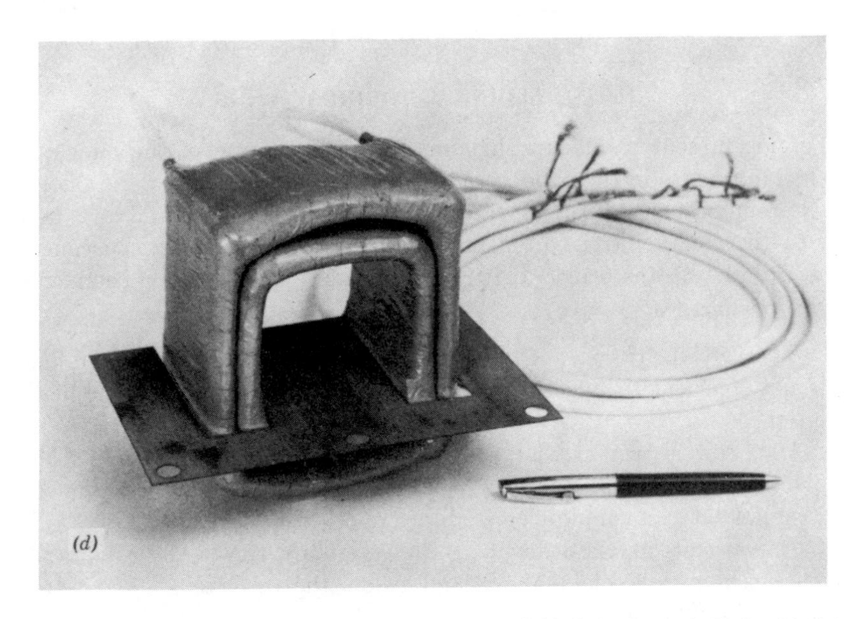

Figure 8.15*d* Nested-box-shielded coils with shielded leads and *E* lamination.

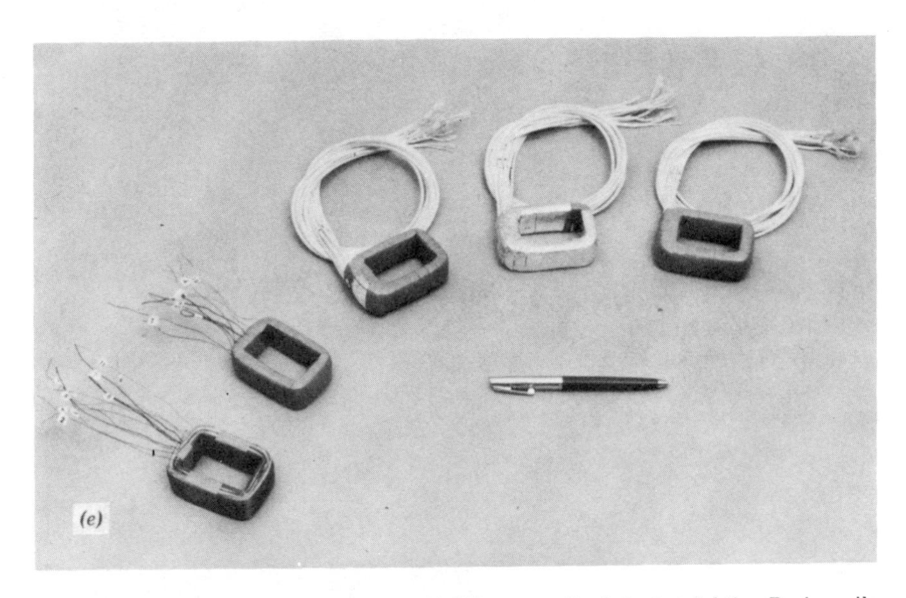

Figure 8.15*e* Steps in double-box-shielding a coil (left to right): Basic coil; coil insulated; coil shielded with leads served; second box shield applied; complete coil ready for laminations.

8.16 SHIELDING A TOROIDAL CORE

Toroids present a difficult shielding problem because of the bunching of shield material in the window. The techniques for treating these shields vary from preformed shield cups to vacuum deposition and are best left to the experts. It is often necessary, in engineering, to experiment with a toroidal transformer. If it requires shields, the following comments may prove helpful:

1. The core itself can act as a coupling element if care is not taken. It is not sufficient to place coils on opposite sides of a toroid to achieve isolation.
2. If a coil is wound that covers the entire core and then a shield is placed over this coil, the core is completely screened out by this coil and cannot act as a coupling capacitance to other coils.
3. If two coils are laid down on the core, only one of the coils needs to be boxed in (a shield above and below the coil) to keep the core from coupling the two coils together.

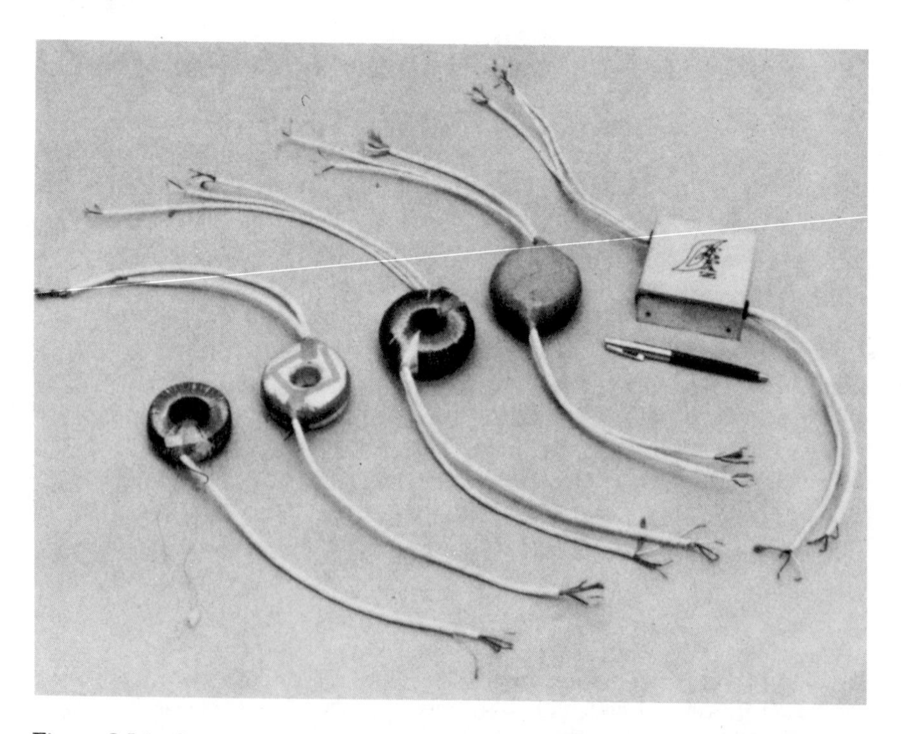

Figure 8.16 Construction steps in triple-shielded toroid. (Photo courtesy of Topaz, Inc., San Diego, Calif.)

Figure 8.16 shows the construction steps in shielding a toroidal transformer.

8.17 BALANCED TRANSFORMER CONSTRUCTION

Coil balancing can reduce circulating current flow by simple cancellation. If the presence of a shield adversely affects other parameters it may be a good compromise to avoid the shield. The following comments are guidelines if balanced construction is used:

1. Use center-tapped coils if possible so that midpoints can be properly referenced.
2. Treat each half of the coil with mirror symmetry referred to this midpoint. This means that points equidistant from the center point have equal capacitances to other points of symmetry in the transformer.
3. Bifilar coil winding can be used. This ensures that equal and opposite voltages referenced to the center point occur together along the coil. This practice is best applied in low-voltage circuits.

8.18 A SPECIAL SHIELDING TECHNIQUE

In small toroids and in high-frequency circuits copper shields are often not satisfactory. A device that is sometimes employed to obtain electrostatic isolation involves two transformers. The coupling circuit acts as the electrostatic shield. This technique is shown in Figure 8.18*a*. The coupling capacitance c_{13} can be made quite small in this arrangement. The disadvantages are obviously cost, transformer efficiencies, size, and so on. If three transformers are used, two shields are provided.

The transformer coupling in Figure 8.18*a* is not free from difficulty. If the coupling circuit is used as a primary shield, the shield's own voltage represents an effective opening in the shield and couples to the next circuit. This can only be avoided by balancing of the coupling circuit's

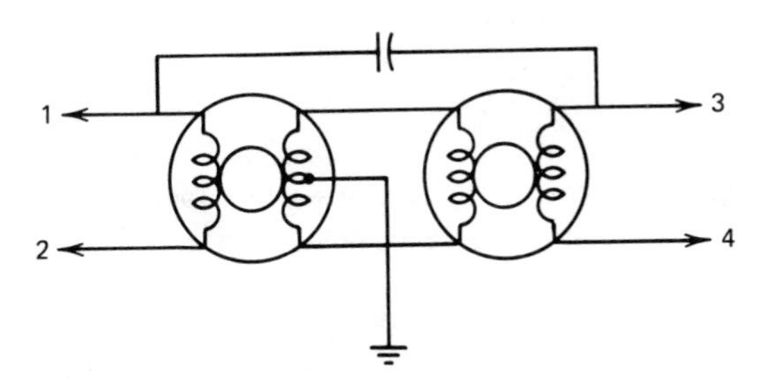

Figure 8.18*a* A coupling circuit used as a shield.

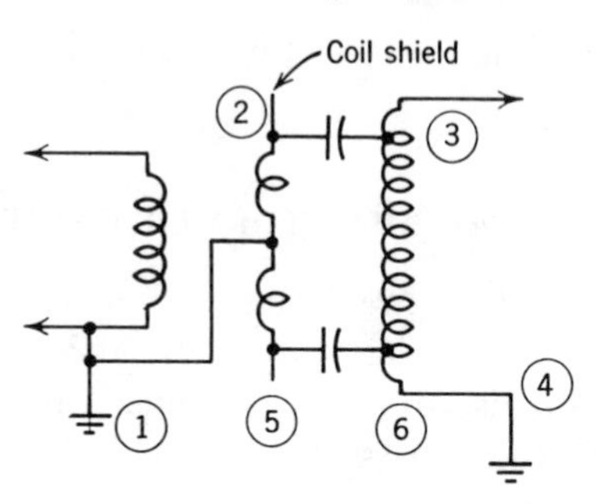

Figure 8.18*b* Equivalent circuit for a coil used as a shield.

center tap and by using a step-down ratio to the coupling circuit. This has the obvious effect of reducing the voltages per turn, which reduces the effective opening in the shield. If this idea is carried to the extreme, a single turn can couple energy and act as a shield at the same time. The single-turn shield is discussed in Section 8.10. The problems indicated above are shown schematically in Figure 8.18*b*. The shield ② has potential differences V_{12} and V_{15}. The current flowing in capacitances c_{23} and c_{56} will cancel if potentials V_{12} and V_{15} are of opposite sign. The current flow will be minimum if the potentials V_{12} and V_{15} are small.

8.19 MEASURING TRANSFORMER MUTUAL CAPACITANCES

The mutual-capacitance values that define shield effectiveness can be measured with the transformer unenergized in the circuit shown in Figure 8.19. The rules for finding the capacitance c_{23} are as follows:

1. Connect all coils, shields, and conductors together that are within the shield ①.
2. Place a signal generator between this tie and the shield ①. Use the shield as the signal return for the signal generator.
3. Connect all other conductors together and place them in series with a resistor R that returns to ①.
4. Any signal appearing across R represents current flow in c_{23}. This can be observed with a voltmeter or an oscilloscope.
5. Use Eq. (1) to calculate the mutual capacitance.

For measurements involving 1 pF or less, a generator frequency of 1 kHz or greater and a resistor value of 10 kΩ or greater are recommended. The capacitance calculation is straightforward. If V_{sig} is the

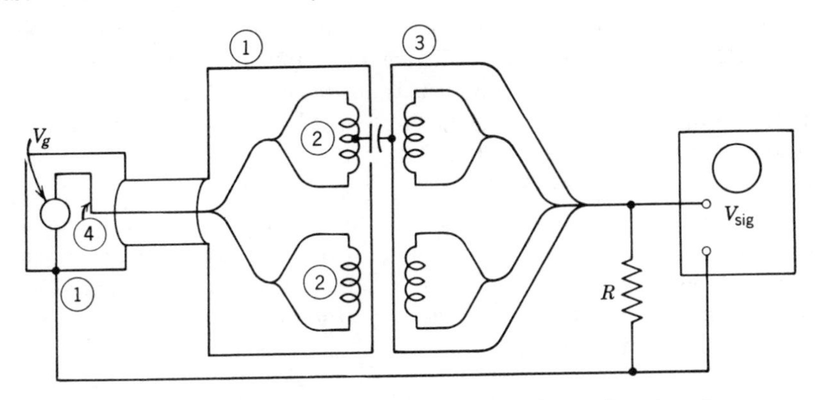

Figure 8.19 Method of measuring mutual capacitance in a transformer.

signal voltage across R, and if V_g is the signal generator voltage, the mutual capacitance in picofarads is given by

$$C = \frac{10^{-12}V_{sig}}{2\pi f V_g} \tag{1}$$

Each shield must be tested separately. For example, to test shield ③, all conductors within this shield must be connected together and driven with respect to the shield as zero reference. The resistor R is then used to sense all currents permitted to flow in the mutual capacitance to all other conductors.

It is imperative that the signal source be completely shielded by ①. Open wires are not satisfactory. If, in Figure 8.19, the signal source has a finite mutual capacitance c_{43}, the measurement will be the sum of c_{43} and c_{23}. Since c_{23} can be quite small, c_{43} must obviously be below this value.

8.20 ACTIVE MEASUREMENTS OF MUTUAL CAPACITANCE

If mutual capacitances are to be measured in an active sense, with the circuits operating, the problem becomes quite complex. In the common-mode rejection processes of differential amplifiers the rejection ratio was also a measure of mutual capacitance, and this was straightforward. Only a few examples exist where the performance and a simple mutual capacitance value are directly related.

If resistors are placed in series with conductors to sense current flow, the circuit malfunctions may negate the measurements. Even if the resistor is tolerated, the measuring device may influence the answer. For these reasons, the active measurement is not recommended.

It is obvious that the performance of a circuit or a system is a measure of those mutual capacitances that affect the performance. In this sense active measurements of key mutual capacitances are automatically made whenever a system or circuit is functioning. This test lumps all effects together, even those that might have a nulling or canceling influence. When possible, worst case conditions (maximum source impedance, longest signal lines, highest signal gain, etc.) should be used to obtain the most information.

9

Rf Processes in Instrumentation

GENERAL

The electromagnetic field enters into instrumentation processes in a number of ways. A detailed and analytic treatment is outside of the scope of this chapter. A discussion of some of the processes can, however, provide some insight into this troublesome area.

The rf problem in instrumentation is not a controlled one. Conductors act as waveguides when they are designed as low-frequency shields; instruments act as mismatched terminations because here again the design criteria did not involve rf. Because of this lack of order, detailed explanations are not as important as the recognition of basic phenomena.

9.1 RADIATION ENERGY

The electromagnetic field discussed in nonmathematical terms provides some insight into the process of radiation. The idea of radiated energy free from circuits or conductors is intuitively difficult to grasp, although we accept forms such as sunlight or radio transmission without question. When a potential difference is suddenly impressed on a group of conductors, charges flow out onto the conductors. The resulting potential distribution causes an electrostatic field and the current causes a magnetic field. (The current must be considered as flowing in the distributed capacitances of the circuit as well as in the conductors themselves.) This combined field is called an electromagnetic field.

The flow of current can not instantaneously propagate over the entire circuit, and as a result, a period of time must elapse before the electric and magnetic fields reaches equilibrium. If the applied potential is changed to a new value at any time a new field must be established and the old field must decay.

The electric and magnetic fields extend into and beyond the space between conductors. It is correct to say that the energy of the system resides in these fields. Just as the potentials and currents propagate at a finite rate along the conductors, the fields propagate into space at the speed of light. When the driving potential is removed, the field at any fixed point must take a finite time to return to zero. The energy stored in the field can either return to the circuit in this time or escape as radiation.

A sinusoidal drive gives some insight into this radiation process. Any developed magnetic field is in phase with the driving current. When the magnetic field energy that is built up is returned to the circuit, not all of it returns in phase with the current causing the field. An out-of-phase component results because of the time required for the energy to travel away from the circuit and return. This out-of-phase component cannot re-enter the circuit and it ends up as propagated or radiated power.

9.2 CONTROLLED RF PATHS

Rf energy tends to follow conductors. In the sense that the energy travels in the space between these conductors, the energy is still electromagnetic in nature even though it is steered by the conductors.

The familiar transmission line, or two conductors running in parallel, is an example of an rf pathway. Another example is a conductor running in parallel with the earth plane. Electromagnetic energy can also propagate between parallel metal surfaces or inside a rectangular enclosure. All of these transmission mechanisms are well understood and many important applications use these processes.

The needs of instrumentation involve conductors in very typical transmission-line configurations. This is almost unavoidable and is a source of most difficulty. Energy will not propagate along these paths unless it can enter into the system. Points of entry are synonymous with points of discontinuity. If a system can radiate energy at such a point it can also pick up external energy at this same point. Once captured, it is brought by conductors to circuit points as contamination or interference.

An analogy with lightning may be of help. Lightning can exist between clouds and not be referenced to earth or to any conductors. If lightning is captured by conductors on the earth, it follows these conductors as the simplest path. The same thing occurs with rf energy. It can travel in space not referenced to earth or any of its conductors. If the rf is captured on the earth, it follows these conductors into the instrumentation.

9.3 THE TRANSMISSION LINE

Two conductors in parallel constitute the simplest transmission line. If the lines have an inductance L per unit length and a capacitance C per unit length, a potential wave can travel along this line in either direction with velocity $v = 1 \sqrt{LC}$. If a potential V is impressed at a time $t = 0$ between two long conductors, a voltage wave will travel along the conductors for a distance z in a time equal to z/v. The current that flows from the driving source is determined by the characteristic impedance $Z_0 = \sqrt{L/C}$. If this impedance is 100 Ω, a 10-V potential causes a 100-mA current.

The transmission line can be stopped or terminated at any point. If a resistor $R = Z_0$ is placed at this point, signals reaching this point cannot tell the difference between an infinite line and this resistor. For this reason a properly terminated transmission line looks like an infinite line.

At any point on the line, the current and voltage relationship are defined by the characteristic impedance. If the line is opened at a point, the current must obviously be zero at this point. This seems to result in a contradiction. An explanation of what occurs can be found mathematically, but the following arguments can be useful.

A transmission line can handle any number of signals traveling in both directions. The boundary conditions at a termination require that the sum of signals at this point satisfy the current-voltage relationship of the termination impedance. For example, when a forward wave reaches the open circuit in the open-line problem, a reflected wave must be developed that cancels the current flow at this point. From this time on, two voltage waves are traveling in opposite directions along the transmission line.

If the line is shorted at a point, the reflected signal must cancel the voltage. This is a negative voltage traveling back toward the source. The current direction for this signal is the same as the initial positive current I_0 making the total current flow $2I_0$. When this reverse signal reaches the source, the current must increase to $3I_0$ to cancel the returning signal and reestablish the correct voltage V. This step buildup of current continues as one would expect in a short circuit.

Viewed on a sinusoidal basis, the impedance of a matched line is Z_0 at all frequencies. For a termination R other than Z_0 the impedance is complex and is dependent on line length. An analysis involving impedance concepts and sinusoids is not simply applicable to the analysis of step-function response. The idea that a voltage wave front traveling at a velocity v reflects at a point of discontinuity gives the simplest physical picture of transmission-line phenomena.

9.4 TRANSMISSION LINES IN INSTRUMENTATION

If two conductors run parallel to the earth, a transmission line is formed. In this arrangement, energy stored in the electromagnetic field of the conductors is modified by the presence of the earth plane. The energy distribution is also dependent on the current flow that may take place in the earth. If the lines are driven symmetrically with respect to the earth, the earth current will be zero.

A transmission line can consist of a single conductor driven with respect to its shield or a balanced drive between two conductors within a grounded shield. In fact, any array of parallel or nearly parallel conductors serves the purpose. In instrumentation processes, the bulk of input cables are arrays of such parallel conductors. They are usually carried in parallel with the earth plane and make excellent pathways for electromagnetic energy to follow.

9.5 RF COMMENTS

One common misconception in rf involves the idea that the rf is "in" the conductors. It is more meaningful to state that the energy is stored in the space between conductors. The conductors act as hangers or runners for guiding the rf energy, but the energy exists mainly in the free space, not in the conductors.

Much as a pressure wave would go down a long tube, rf energy travels between two conductors. When the pressure reaches a reflecting discontinuity a wave is sent back, modifying the tube pressure on its return. One does not think of the pressure wave as only being at the walls of the tube. It is intuitively obvious to consider the pressure filling the volume, although it may be sensed at the wall.

9.6 WAVEGUIDES

Electromagnetic energy can be supported within a single long cylindrical metallic conductor. This conductor can take the form of a cable tray, a long array of rack cabinets, an aircraft fuselage, an access tunnel, etc. The conditions for supporting energy are closely related to the dimensions of the enclosure. A minimum frequency exists for this waveguide action. Below this frequency any energy that exists is attenuated rather than propagated. The following table gives the minimum frequency f for several geometries where c is 3×10^{10} cm/sec, the velocity of light.

Parallel conducting planes	$f = \frac{c}{2a}$	a = spacing
Rectangular guide	$f = \frac{c\sqrt{a^2 + b^2}}{2a}$	a, b, dimensions
Circular guide	$f = \frac{0.289c}{a}$	a = radius

Digital signals contain pulses and leading edges that rise and fall in a few nanoseconds. These leading edges can have frequency content above 30 MHz. This type of signal can be propagated in a waveguide. An examination of the equations above shows that the dimensions for supporting 30 MHz between parallel planes is 15 ft.

9.7 RF SHIELDING

The processes for protecting low-level signals are often opposite to that required to eliminate rf interference. These two requirements can clash, particularly where bandwidths above 100 kHz are a consideration.

As previously discussed, rf energy reflects at points of discontinuity. When a shield is grounded or earthed this represents a reflection point for energy using the shield as a transmission line. If this shield is signal grounded once, then a second ground that may be required for rf elimination is undesirable according to the rules established for signal shielding.

We should treat rf shielding as a separate process so that signal shielding is not violated. If an rf shield must be connected to ground at many points, it should be a separate shield over the signal shield. Often such an rf shield can enclose many signal cables at one time. This solution may involve the use of a cable tray or raceways or conduit.

The rf energy usually enters a signal process at a point of discontinuity such as at the ends of a signal run. Precautions taken along a run may not prove to be effective, and it can be impractical to provide rf shielding at the point of pickup. It is often simpler to accept the rf pickup and then use rf filtering by passive elements at critical points along the signal path.

It is important to note that a filter may not look like a matching impedance to the transmission line. It may only provide a point of reflection so that energy will not pass that point. An improperly placed reflection point can act to redistribute the problem, not solve the problem. A filter placed at a point of observation reduces the rf at this point. Rf at other points can still exist, and if this causes no difficulty the problem has effectively been solved.

9.8 SKIN EFFECT

Current flowing in a conductor establishes a magnetic field. This magnetic field causes a distribution of current flow that varies as a function of frequency. The current tends to flow on the surface of a conductor and thus we have the term skin effect.

For a solid, round copper wire the effective resistance can be expressed as factor times the dc resistance. Specifically

$$R_{ac} = k\sqrt{f}\,R_{dc} \tag{1}$$

where f is frequency in megahertz, and k has the following values for various conductor sizes:

Wire size	k value
No. 22	6.86
18	10.9
14	17.6
10	27.6
8	34.8
6	47.9
4	55.5
2	69.8
0	88.0
00	99.0
0000	124.5

Thus, at 1 MHz, an "0" wire looks like a No. 19 wire at dc.

9.9 GROUND CONDUCTORS

It is of interest to note the inductance of a segment of straight conductor. (Inductance only has meaning for a closed circuit, but the contribution from each segment can be considered separately.) A first approximation (in microhenries) for inductance for a long conductor at high frequencies is

$$L = .0051l \left[2.30 \log_{10} \frac{4l}{d} - 1 \right] \tag{1}$$

where l is the conductor length and d is the diameter in inches. For 1000 ft of No. 2 wire this is about 572 μH. At 60 Hz, this is 219 mΩ. The resistance of No. 2 wire is 156 mΩ/1000 ft. The inductance L in (1) and the conductor diameter are logarithmically related. For this reason low values of inductance are not easily obtained by adding copper cross-section. In fact, for even very large conductors, the reactive component of impedance dominates above a few hundred cycles.

It thus appears impractical to "short together" two remote points by connecting them with copper conductors. Such a connection provides a path for many amperes to flow and the impedance will support a significant potential difference at high frequency. The table below lists a range of conductor sizes and their resistance and inductance for 1000-foot lengths.

Table 9.1

Wire Gage Number	Diameter in Inches	Milliohms/1000 Ft	Inductance/1000 Ft in Microhenries
0000	0.460	49.0	540
00	0.364	77.9	—
0	0.325	98.3	—
2	0.258	156	572
4	0.204	248	—
8	0.128	628	—
10	0.101	999	616
14	0.068	2,525	—
18	0.040	6,385	—
22	0.025	16,140	636

9.10 SHIELDED ENCLOSURES

Commercial enclosures are available where the rf levels are kept quite low. These enclosures are ideally suited for testing where rf interference would present a problem. Since power entering the enclosure is a source of rf contamination, it must be carefully filtered at its entry point.

An rf enclosure can be violated if contaminating currents are permitted to flow in the walls of the enclosure. These currents create an electromagnetic field that can be sensed inside the enclosure. For this reason, an rf enclosure should be single-point grounded. Any power entrance (a ground itself) should be at this same physical point.

The problems of floating a large screened environment to maintain a single-point ground are obvious. Many commercial enclosures rely on double walls in their construction so that the outer enclosure can by grounded by its environment and still not cause a field within the inner enclosure.

9.11 RF-PICKUP ELIMINATION

The key word in rf elimination of radiation or pickup is loop area. To implement this idea, signal leads and their return conductors should be kept close together. When this is done, the external magnetic field is reduced along with any radiation. Conversely, if a signal lead and its return conductor are wired close together they present a small flux-capture cross section for any external field. This reduces radiation pickup in the circuit.

In digital systems, it is impractical to keep all return paths next to signal leads. The best practice involves shortest-distance wiring (reduced inductance) and a common return-conducting plane near the conductors. This ground plane can be sheet copper placed just over or under the wiring plane.

In specific cases, such as with timing pulses, the two conductors carrying the signal can be twisted together along a run. This may be an additional procedure over and above the use of a conducting plane as described above.

The copper sheet as an rf ground plane may exist in several segments, depending on the dimensions of the circuitry involved. Because of inductive effects and because of physical dimensions, an ohmic tie between conducting sheets may not guarantee that they will act as a single ground plane. For this problem, past experience and common sense play an important role in designing a practical system. It is obvious to state that the faster the pulses and the greater the operating speed, the more difficult it is to contain the radiated energies so that information is properly channeled by the conductors, not by any radiated electromagnetic field.

Several errors are often made in using ground planes in digital circuitry. The idea is to provide a ground return path for each signal wire near the wire. If the ground path is much longer than the signal path, the ground is ineffective. Common errors include:

1. Single-point connections to the ground plane at edge connectors. Multiple connections are preferred, distributed along the connector.

2. No provision for ground plane extension on motherboards. Signals processed between digital circuit cards are as important as signals processed on a single card.

3. Omission of ground leads along with digital lines in cables. Several ground conductors or a shield surrounding the signal conductors can be effective.

It is not common practice to terminate most digital signal transmissions because the interconnecting wires are normally quite short. If an appreciable distance is involved, then the signal conductors must be treated as regular transmission lines.

9.12 RF COMMON-MODE SIGNALS

High-frequency interference problems exist in almost all systems. If the instrumentation has sufficient bandwidth, this type of signal can be difficult to eliminate. The 60-Hz problem involved capacitances of 2 pF. At 1 or 2 MHz, the capacitance levels permitting common-mode current flow are correspondingly smaller.

Fortunately, at high frequencies the shunting effect of input cable capacitance tends to remove the source unbalance that dominates the common-mode rejection process. Also the characteristic impedances of various transmission lines rarely get above a few hundred ohms. It is these impedances that define the reflections and distribution of rf energy.

The input shield in a differential amplifier is open ended at the instrument. This is a reflection point for rf energy carried by the shield and earth as a transmission line. To connect this shield to earth at the instrument to eliminate the rf signal violates the shielding rules. A partial solution involves bypassing the shield to earth for rf only at the instrument. If this is done, the previous open-circuit reflection point becomes a short-circuit rf-reflection point. This type of solution may not be as

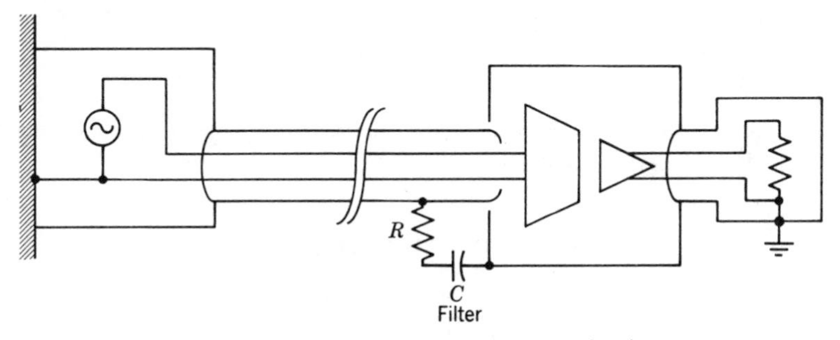

Figure 9.12 Shield bypass for rf reduction.

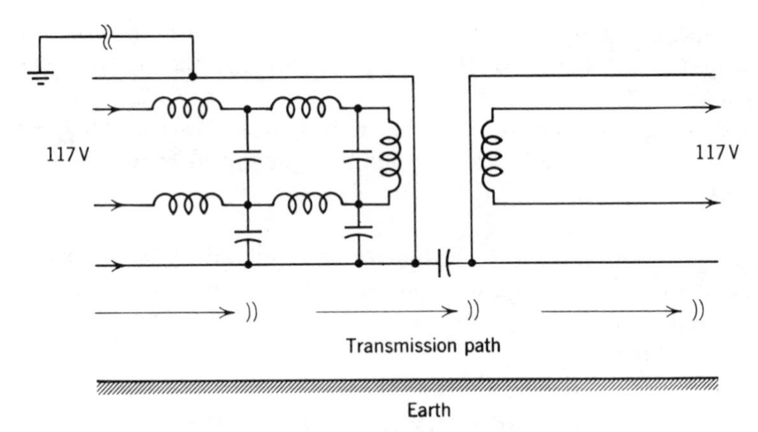

Figure 9.13a Rf traveling along the power cable–earth transmission line.

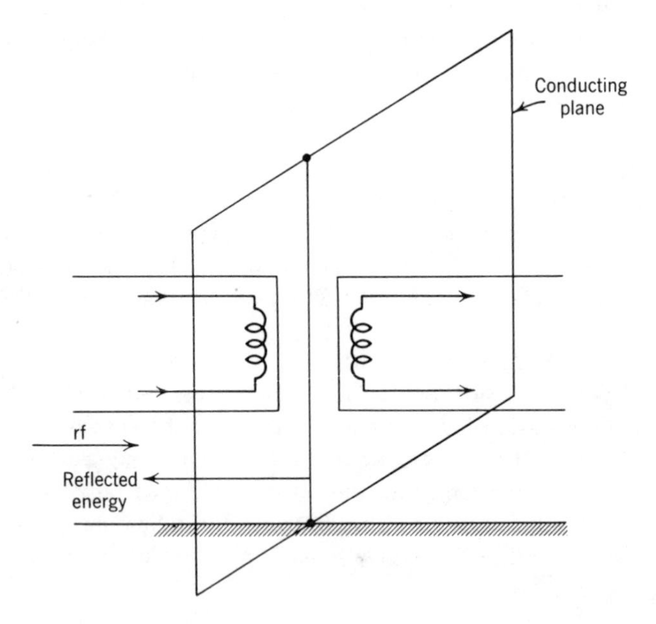

Figure 9.13b The third shield in an isolation transformer.

effective as a second rf shield cover, but it is much less expensive. The filter location is shown in Figure 9.12.

9.13 THE ISOLATION TRANSFORMER

If rf filters are placed across a power entrance they are ineffective in stopping rf energy traveling between the power line and earth. The energy can often, however, be reflected by a short placed across the power cable-earth transmission line. Unfortunately, the geometry that is available does not always permit this on a practical basis. Figure 9.13*a* shows how rf can travel across a transformer with two shields even though a filter exists on the line. The shields may be grounded remotely but this only acts to reflect rf energy at these points.

A third shield that is connected to a conducting wall at the transformer location can effectively reflect rf energy. This arrangement is shown in Figure 9.13*b*. Here, rf energy entering from the left is reflected back to the left and does not enter the environment on the right.

9.14 LONG SIGNAL LINES

Frequency response must be considered when signal lines are long. See Section 9.18. The problem becomes particularly important when the signal lines approach a quarter wavelength. At this length, line terminations must be considered. Because dielectrics must be used in cable construction, the velocity of propagation is about one-half to one-third the velocity of light. At 25 kHz, a quarter wavelength is about 3000 feet. If the signal transmission is to be held to 1% at this upper frequency, then line terminations may have to be considered on shorter runs.

Signal transmission usually does not lend itself to line matching. Fractional or partial matching may prove useful. Here a resistance other than the characteristic value is inserted in series with the signal source or across the terminating point. These values are best found empirically while monitoring signal transmission. Square-wave signals at a low and high frequency can be used as a convenient criteria. When the square-wave leading edge is optimum and when any sag phenomenon is eliminated the line is adjusted for best performance. Figure 9.14 shows the waveforms that might be encountered. Typical high-frequency square waves might be 2 kHz and the low-frequency values might be 20 Hz.

The discussion above is meaningful for many transducer types. A piezoelectric transducer looks like a capacitor, and this certainly does

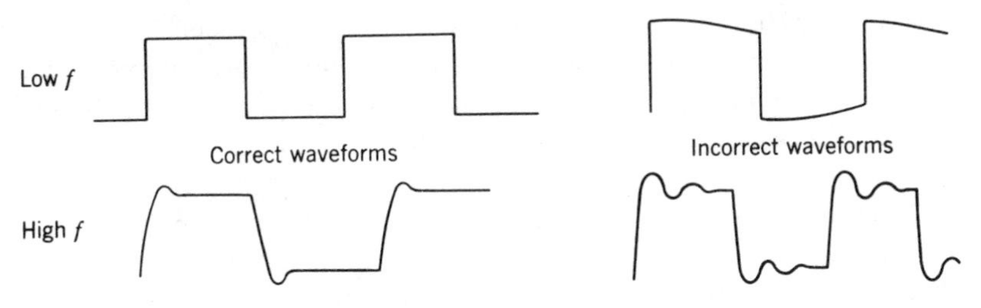

Figure 9.14 Square-wave response of an adjusted line.

not match to a transmission line. Strain-gage elements are usually not selected for line-matching qualities. In many cases, some simple line modification can optimize the frequency response, but if not considered, the response can be unfavorable.

9.15 DRIVEN LINES FOR CAPACITANCE REDUCTION

Double-shielded conductors can be used when critical capacitances must be reduced electronically. The arrangement is shown in Figure 9.15. When the signal kE is returned to the inner shield conductor ①, the signal voltage must only supply the charge between ③ and ①, which is $Q = C(E - kE)$. If k approaches unity, then Q approaches zero. The new capacitance seen by the input signal is $C(1 - k)$.

If the input lines are short, the capacitances affecting the feedback can be treated as lumped parameters. The complex nature of the gain factor k and these capacitances can be analyzed to determine conditions for stability.

If the lines are long enough to approach an eighth wavelength, the lines must be considered transmission lines, not simple capacitances. To obtain a clearer picture of the phenomena, it is meaningful to apply a step voltage E to the system. If we assume E has a source impedance R_1, then the voltage wave that travels down the line towards the amplifier is $ER_L/(R_L + R_1)$ where R_L is the characteristic impedance of the line. This reduced signal reaches the amplifier and $kER_L/(R_L + R_1)$ returns down the inner shield towards the source. When this signal is returned, the source is further unburdened and a higher voltage, still less than E, travels down the line towards the amplifier. This process continues until eventually the signal reaching the amplifier is nearly E and the charge on the line is supplied mainly by the amplifier, not by the source.

If the input capacitance of Figure 9.15 is to be reduced at a frequency f, the complex nature of k must be considered at this frequency. If k has a quadrature component a resistive term will appear in shunt across the input. The effect caused by the transmission line is more difficult to analyze, but it too results in an input resistive component. An analysis involving both effects is quite complex so the following rules will suffice to indicate the limits of this technique.

1. Keep the phase shift in kE below 10° at any frequency where C is to be reduced.
2. Deal only with input lines whose length is less than one-eighth wavelength, assuming a signal velocity of $c/2$ where c is the velocity of light.

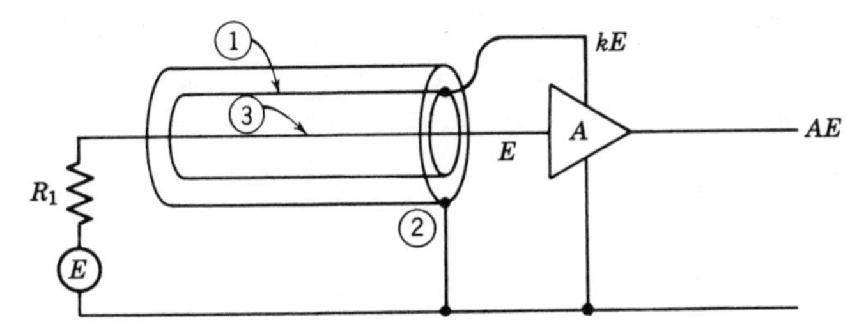

Figure 9.15 A driven shield arrangement.

9.16 TRANSMISSION-LINE EQUATIONS

Formulas providing values for the characteristic impedance of various conductor configurations are given below. In each case, the earth is assumed to be a perfect conducting plane. Let l be the conductor spacing, d the conductor diameter and h the distance between the conductors and the earth plane. Further, let

$$\log \frac{4h}{d} = \alpha, \log \frac{2l}{d} = \beta, \log \frac{2h}{l} = \lambda$$

$$Z_0 = 138\alpha$$

$$Z_0 = 276\beta$$

$$Z_0 = 138\beta \frac{(2\lambda + \beta)}{\lambda + \beta}$$

$$Z_0 = 138 \left[\alpha - \frac{(0.301 + 2\lambda)^2}{3\lambda + \alpha - 0.146} \right]$$

In critical applications where transmission lines must be used to reduce pickup or radiation from adjacent signals, printed circuit tracks can be effectively used. The ground plane alone is often not adequate as it is a common conductor to all signals. Geometries that are possible include parallel tracks on one or two sides of the circuit card, guard tracks carrying ground, or pairs of tracks arranged on each side of the circuit card with diagonal elements in parallel. These configurations provide varying degrees of isolation at very low cost.

9.17 SIGNAL CABLES

A wide range of cables is made in the United States for use by industry. These cables range in outside diameter from 0.008 in. to several inches, depending on application. Cables are available in multiconductor configurations with many shielding and insulation arrangements. The combinations available are too numerous to list and the types and arrangements are constantly undergoing changes.

Many cables are assigned a military code number. This code includes a letter prefix RG/U and a simple numerical value. Most manufacturers refer to this military code number when applicable in their catalogs, and this provides a convenient cross-reference guide. As time goes by, many cable types are obsoleted and replacement items are designed and catalogued. The military specifications covering cables and insulation are referenced to MIL-C-17.

The majority of shielded cables are designed for applications not involving instrumentation. These include rf transmission, waveguides, high-voltage transmission, hazardous environments, television transmission, etc. These cable types can often serve in a variety of secondary applications. However, when a large amount of cable is involved, special manufacturing runs are often made to effect a savings and optimize performance.

9.18 CABLE FREQUENCY vs AMPLITUDE RESPONSE

One of the most important cable parameters is attenuation above a few MHz. These losses occur in varying degrees depending on the cable design, that is, dielectrics, diameter, conductor size, and so on. For instrumentation, the problems are usually below 1 MHz and line losses of this type are not a difficulty.

Below 1 MHz, the unterminated frequency response of a cable must be considered. Unfortunately, these data are rarely made available to the engineer. If they were, he might be shocked. For example, it is possible for there to be a 2-dB drop in frequency-vs-amplitude response at 10 kHz in a 100-foot cable run. Many a wideband system has been built where the cable defines the frequency-vs-amplitude response and this is unknown to the engineer. The individual parts are verified for response, but it is often difficult to verify the response over the finished system. The result is data modified in an unknown manner—certainly not a desirable condition. See comment in Section 9.14.

Since the unterminated frequency response can vary widely depending on cable length, cable type, shield connection, source impedance, and so on, it is not possible to solve this problem in a general equation for presentation. The important thing to appreciate is that a cable is a distributed-parameter system associated with lumped-parameter components. If accuracy in phase or amplitude is required, the cable must be considered along with the instrumentation.

9.19 CABLE-SHIELD EFFECTIVENESS

Many shielded conductors are commercially available where shield effectiveness is only 70 to 80%. This percentage is usually a rating of the surface area actually covered by shield conductors. This coverage can only be meaningful when the mutual capacitance effects to other conductors are considered. Mutual capacitances of 0.1 pF/ft are not uncommon when the shield effectiveness is only 70 to 80%.

One method of tighter shielding involves the use of aluminum or copper foil wrapped around the conductors. This foil is usually not mechanically acceptable as a shield by itself and a drain wire in electrical contact with the foil must be carried along the cable length. The foil technique gives a nearly perfect covering, thus reducing the mutual capacitance effects to negligible levels.

Another method of providing more effective shielding involves the parallel use of two shields. This unfortunately is expensive, but if the shielding is available it can be used for this purpose. The second shield is often better used for rf elimination, as indicated in Section 9.7, even though the potential on this outer shield can couple into the signal conductors via the mutual capacitance.

9.20 LOW-NOISE CABLE

Cables can be excellent transducers by themselves. Mechanical motions produce electrical signals much larger than the expected signals. Manufacturers of cable have devised techniques for reducing these effects. The procedure usually involves coating or impregnating the dielectric with conducting material so that charges are not accumulated on the dielectrics surface when the cable is deformed. Although this low-noise cable is usually offered as simple coax, it is available in twinax or in other configurations on special order.

10

The Earth Plane

10.1 GENERAL

The earth enters into the instrumentation process in many ways. It is a footing under facilities and test structures, it carries power currents, and it is used as a reference of zero potential. This wide use is unavoidable and it causes its problems. Techniques that can be applied to instrumentation to accept this usage are discussed in previous chapters. It is still important to consider the earth as a separate element so that instrumentation can be made safe and free from undue power coupling.

10.2 UNITS OF RESISTIVITY

The property of a material to conduct electricity is measured in units of resistivity. If a material has a high resistivity it is a poor conductor. The resistance R of a conductor is given by the statement

$$R = \frac{\rho l}{A} \tag{1}$$

where ρ is the resistivity, l is the conductor length, and A is the cross-sectional area. For (1) to be correct ρ must have units of resistance $\times$ length. Resistivity ρ is usually given in units of ohm-centimeters, abbreviated Ω-cm.

For copper, ρ is 1.72×10^{-6} Ω-cm. This means that a piece of wire with a 1-cm cross section and 1000 cm long will have a resistance of 1.72×10^{-3} Ω.

10.3 TYPICAL SOIL RESISTIVITIES

The earth resistivity is dependent on many factors. For example, moisture, soil chemistry, and temperature all affect the value. Earth resistivity in unprepared areas is usually greater than 500 Ω-cm. Sand and gravel areas can run as high as 10,000 Ω-cm, and rocky areas or areas of shale can even be higher.

The simplest soil-resistivity measurement technique involves a four-terminal method. Here four insulated conductors make earth contact at the bottom of four holes. These holes should be evenly spaced along a straight line so that the hole diameters and hole depth are small compared to the hole spacing. When a current is sent through the outer two conductors, the potential between the two center conductors yields a resistance value R. The soil resistivity is then

$$\rho = 2\pi a R \tag{1}$$

where a is the conductor spacing.

10.4 RESISTANCE AND CAPACITANCE ANALOGY

In a homogenous conductive medium, the current flow between conductors follows the same pattern as the field E in a capacitor. This fact allows electrostatics to yield resistance equations. If C is the capacitance between two conductors in a vacuum, then the resistance between conductors is

$$R = \frac{\rho}{4\pi C} \tag{1}$$

where ρ is the resistivity of the medium.

In most calculations the earth acts as a half-plane of symmetry. Resistance values double when the earth's surface excludes one-half of the conductive paths. The capacitance calculation must therefore include the conductor and its image above the earth.

10.5 THE DRIVEN ROD

The resistance to the earth for a driven rod depends on its diameter D and on its length L. Using methods described above,

$$R = \frac{\rho}{2\pi L} \log_e \frac{4L}{D} \tag{1}$$

For a 1-in. diameter rod buried 20 ft deep this value turns out to be 18 Ω for ρ = 10,000 Ω-cm. For 6-in. diameter rod buried 100 ft the resistance drops to 3.4 Ω. It is thus apparent that very low-value resistance connections to the earth are difficult to make.

Rods driven into the earth in arrays help to reduce this resistance. For four rods, the improvement is usually not greater than 2.5 times the single rod value. For eight rods at a distance of greater than 5 ft, the improvement is a factor of 5. As a further example of resistance values, twenty $\frac{1}{2}$-in. rods buried to a depth of 8 ft and having a spacing of 8 ft provide 4 Ω resistance to the earth for ρ = 10,000 Ω-cm.

Conductors carried along the surface of the earth also make good connections to the earth. A 150-ft run of $\frac{1}{2}$-in. conductor buried 2 ft below the surface yields a theoretical 4-Ω resistance to earth. For this reason, various star arrangements from a central point make practical earth connections. As an example, an 8-arm star of 1-in. conductor buried at a depth of 3 ft yields a 1-Ω theoretical resistance to the earth if the resistivity is 10,000 Ω-cm.

10.6 UTILITY PRACTICE VS LIGHTNING

The current levels in a typical lightning stroke can exceed 20,000 A. It is thus obvious that a low resistance path to earth is critical if the potentials that arise are to be controlled. For this reason utility systems are designed with particular attention paid to tower footing resistances and earth connections around distribution systems. It is unsafe to isolate power distribution from earth and therefore it is used as a neutral conductor, and as a reference point along the distribution path. The National Safety Code requires neutral connections to earth at prescribed intervals along the transmission line.

10.7 UTILITY PRACTICES

Utility currents flowing in the ground plane can vary considerably. Typical values range from 2 to 10 A depending on the nature of the unbalanced loads and the power levels involved.

This earth current tends to concentrate under the transmission line. This results because the energy stored in the resulting magnetic field tends to be a minimum value. It is thus apparent that instrumentation processes should not parallel or run across a transmission line as the earth potential gradients and magnetic fields are apt to be highest along this route.

The earth-current distribution in practice varies with the seasons due to soil conditions. Underground water levels, for example, play an important role. In some cases, other conducting cables or conduits collect the current and distribute it where it is unwanted. To control the ground-current flow, the power distribution can be remotely located and preferably run at right angles to any long signal runs. A distance of one-half mile is recommended if this distance is available when installation plans are being made.

Another practice often used provides a known earth path for the neutral current to follow. A group of conductors can be buried along the path of the transmission line to "collect" the current and insure its flow in a controlled manner. This practice is advisable where soil conditions may be particularly bad during long dry periods. Conductors that inadvertently collect power currents and place them near critical areas must be avoided. A delta connection should be used at least once in a power-entrance complex. This provides a path for higher-harmonic current flow created by nonlinear loads, etc. If the delta connection is not made, then these currents must flow in power conductors resulting in an increase in electrostatic and magnetic fields near these conductors.

10.8 A PROBLEM IN EARTHING MANY CONDUCTORS

It may be desirable to earth several conductors for lightning protection. If these conductors are ohmically tied together, then earth currents can use these conductors in preference to the earth itself. This current flow in these conductors causes a magnetic field, which can be undesirable in the vicinity of instrumentation. The correct practice involves a single-point earth connection so that earth currents will not accept this path. If this practice is unacceptable for lightning protection, then above-earth connections may have to be gapped or redistributed to avoid any lightning hazard.

10.9 GROUND BUS

The practices indicated in previous chapters did not suggest elaborate earthing procedures. Use of an earth or ground bus was not discussed. The problem that usually occurs when a common bus is used involves the currents that flow from each connection to the bus. These currents can be 60 Hz or rf "hash" and the bus can never be low enough in impedance to "earth" everything to that "true" zero of potential that is thought to exist. Instead, each user of the bus sees a signal along the bus caused by the drain currents of all other users. If the bus were used to drain only controlled currents, it might serve a useful purpose.

Because this is rarely possible, instrumentation design that requires this type of earth connection should be considered inadequate.

If a ground bus is required for safety purposes, its ohmic connection to earth can be enhanced by salting and wetting the soil around the buried conductor. This procedure drops ρ and increases the effective conductor diameter. Resistances below 1.0 Ω can be realized with this technique. Using this method resistance below 5 Ω should be a safety objective. It should be pointed out again that above a few hundred hertz impedances will not be this low in value. The ground bus is protection against the effects of equipment faulting and lightning, but it is not the source of zero potential for the solution to instrumentation processes.

10.10 FLOW OF NEUTRAL CURRENT

Safety codes require grounding of the power neutral along the transmission line and at a distribution point. If the power neutral is grounded at an instrumentation complex, neutral currents will flow in the earth and add to the ground difference of potential. The neutral should be carried as a separate conductor to avoid this source of difficulty. All metal enclosures and housings should still be earthed and bonded together but no current should be permitted to flow in these connections. Currents will flow if accidental shorts or malfunctions occur and this is just the safety factor required. Figure 10.10 shows such an arrangement.

The National Electrical Code excludes a neutral earth tie beyond the main service disconnect. If this rule is followed, earth currents can still be brought near the instrumentation, which may not be desirable. The intent of Figure 10.10 is to keep the earth-current flow far removed

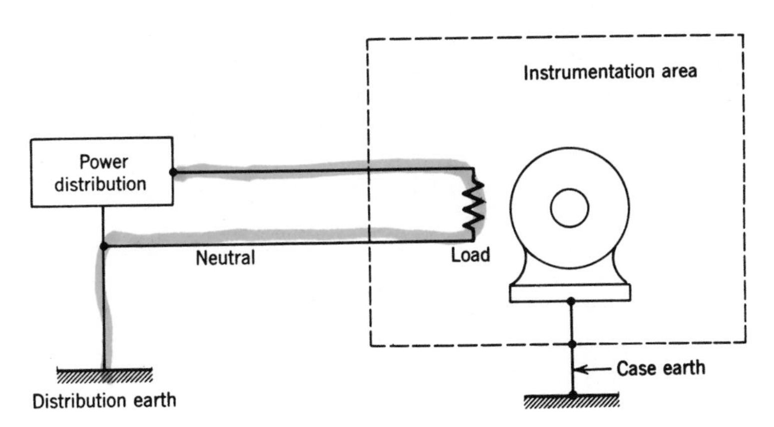

Figure 10.10 Power neutral treatment.

from the installation, and this requirement needs to be intepreted in terms of service location, existing installations, etc.

10.11 FLOATING THE INSTRUMENT RACKS

Commercially available instrumentation often operates with a permanent connection between cabinetry and circuitry. This is simply a statement that the cabinetry is used as a shield for the signals being processed. When this instrumentation is mounted in a rack, the rack defines a common shield potential for all equipment so mounted. This creates no difficulty as long as this common reference potential is the correct one for all signals being processed.

The rack potential is usually defined by the maze of external connections made to it. These can include building earths, conduit connections, adjacent racks, signal connections, intentional earth connections, reactive ties via power transformers, etc. Good practice suggests that it is wise to insulate the rack from the obvious ties such as building earths and conduit connections so that the rack can be ohmically connected to a potential most favorable to the instrumentation processes. The power-line-reactive connection can be broken up so that current flow is defined in a favorable manner. The process is discussed in earlier chapters. If the isolated rack is connected to the average signal process potential, common-mode difficulty can usually be reduced. This may not be true if the signals are simultaneously derived from widely separated points. This ohmic connection should be made through a heavy conductor, but again this represents an inductance above a few hundred cycles.

Index